Telecommunications Wiring

Third Edition

Telecommunication Wiring

Third Edition

Clyde N. Herrick

Pearson

Boston • Columbus • New York • San Francisco • Amsterdam • Cape Town
Dubai • London • Madrid • Milan • Munich • Paris • Montreal • Toronto • Delhi • Mexico City
São Paulo • Sydney • Hong Kong • Seoul • Singapore • Taipei • Tokyo

Library of Congress Cataloging-in-Publication Data is on file.

Acquisitions Editor: *Bernard Goodwin*
Production Editor: *Rose Kernan*
Cover Designer: *Nina Scuderi*
Cover Design Director: *Jerry Votta*
Manufacturing Manager: *Alexis R. Heydt*
Marketing Manager: *Bryan Gambrel*

For information about buying this title in bulk quantities, or for special sales opportunities (which may include electronic versions; custom cover designs; and content particular to your business, training goals, marketing focus, or branding interests), please contact our corporate sales department at corpsales@pearsoned.com or (800) 382-3419.

For government sales inquiries, please contact governmentsales@pearsoned.com.

For questions about sales outside the United States, please contact intlcs@pearson.com.

Product names mentioned herein are the trademarks or registered trademarks of their respective owners.

ISBN-13: 978-0-13-028696-3
ISBN-10: 0-13-028696-6

61 2024

To Gerry, my helpmate.

Contents

Chapter 8   **Planning the Wiring Installation**   **157**

Preface

$\mathbf{T}$he third edition of *Telecommunication Wiring* has been expanded to include NEC Fire Code requirements for communication wiring, Digital Subscriber Line (DSL) technology, and current wiring and network technology. The data communication field has changed rapidly since the publication of the first edition of *Telecommunication Wiring*. Great strides have been made in the data rates and distance capability of copper wire. The interconnectivity of different protocols has improved nationally and internationally.

The expansion of the Internet and technologies that allow data, voice, and video on the same wire will, in many cases, require reexamination of communication planning. The competition demands that every resource, including wiring systems, be utilized to its maximum capacity.

The need for special wiring systems and greater capacity cabling for data communication equipment has created a generation of new job categories in the workplace, such as telecommunication manager, communication wire planner, information system manager, connectivity specialist, and communication wiring specialist. For these and other occupations in the field, there is an obvious need for a text dedicated to the "nuts and bolts" of telecommunication systems and cabling.

Many books have been written on higher level subjects in telecommunication, such as local area networking, designing LANs, and telecommunication systems. However, the cabling and wiring sections of such books seldom offer any practical information for those involved in designing, installing, testing, or updating wiring systems—which are all critical to the operation of any telecommunication system. The cabling should be treated as a "dynamic source" rather than a static one. These cabling systems, whether a single coax or a complete wiring plant, should be treated as a major support subsystem.

Management will find helpful the discussion of the importance of having a complete inventory of installed cable and wiring runs to determine "in place capacity" versus "in place used capacity." The chapter on task management will assist managers in giving direction and leadership to the installation team, the maintenance team, and upper management in preparing the proposal for and evaluation of the finished product.

Telecommunication cable installers, planners, managers, and audit teams should find useful the discussion of standardization in setting up methods for identifying and labeling. This topic will be particularly helpful if the system has gone through several installs without a set of universal standards. These suggested standards should be a help in establishing corporate labeling standards for cabling, patch panels, wiring closets, floor locations, and equipment.

Wiring specialists and telecommunication planners/designers should find the discussion of cabling systems, supports and test hardware, proper installation techniques, and wire and fiber characteristics useful in the planning of a cabling and wiring system. The chapter on planning the wiring installation offers the wiring specialist guidelines for planning, installing, and testing the cabling system. Finally, the chapter on premise wiring should aid system planners in developing a wiring plan and help cabling and installation specialists in selecting hardware for the installation.

The book attempts to establish a reference point from which logical decisions about designing a cabling system, selecting the media type, writing the job proposal, documenting the system, and establishing a maintenance facility can be completed—with the understanding that every company has unique telecommunication needs and that every wiring system will be different.

While the text includes certain trade names and trademarked items, this is not to be taken as an endorsement of any particular product. These examples are included to illustrate some of the more successful products and telecommunication wiring techniques and options on the market today. There are many manufacturers and vendors for most of the items mentioned, and it is the responsibility of the professional to keep abreast of the literature. To this end, names and addresses of many of the periodicals in the field along with some of the vendors mentioned in the text have been included.

The author has attempted to make the revision of *Telecommunication Wiring* as "state of the art" as possible, fully realizing that technology in this field changes daily.

The author wishes to express his appreciation to all the companies and individuals who have supplied information for this book and its revision.

The author would appreciate any suggestions for improving this book, including any new topics in the field that deserve future coverage.

Clyde N. Herrick
e-mail gherrick@cwNET.com

Electrical Characteristics of Wire

1.1 INTRODUCTION

It may not be necessary for the worker who installs wiring systems to understand the electrical properties of electrical conductors and fiber optical cables to install a system correctly. However, his or her understanding of these properties will give a better appreciation of the job to be done, the tools that are to be used, and the results of troubleshooting the system.

It is necessary for the staff person responsible for communications and for the wiring system designer to understand how wiring characteristics affect signal information. The purpose of this text is to assist both the cable installer and the wiring system designer.

A **wiring system** is a form of an electrical circuit. An electrical circuit is comprised of an energy source, an energy transfer medium, and a load. An energy source could be a battery, a generator, an amplifier, a digital computer, or any of the other devices that output energy in the form of a voltage, current, or light. **Energy transfer media** are any of the materials used to transport energy from one place to another. Transfer media include copper wires (conductors), fiber-optic cables, and air (in the case of radiated energy). A **load** in a circuit can be any of many components or devices that receive the energy transferred, such as resistors, lightbulbs, speakers, motors, computer terminals, printers, or personal computers (PCs). To understand better the concept of energy transfer, circuits, sources, and loads, we must introduce the concepts of voltage, current, resistance, power, and energy transfer.

1.2 VOLTAGE IN AN ELECTRIC CIRCUIT

An **electromotive force** (EMF) is a force that tends to move electrical energy. Electromotive force or voltage is conveniently regarded as an attractive or repulsive force on charges. Voltage can be compared to water pressure that causes water flow in a pipe, which is measured in pounds

Figure 1-1   A 9-V battery.

per square inch (psi). A voltage is a difference of potential or EMF that attracts and repels electrons. Another way of thinking of voltage (V) or (E) is as a force or pressure that forces electrons through a circuit. The movement of electrons transfers energy throughout the circuit.

DC voltage (direct current voltage) is the name given to voltage in a circuit in which the current flows in one direction only. DC voltage is either positive or negative. This usually means that it is a value above ground (positive) or below ground (negative).

Ground potential or **reference** is considered to be 0 V. Ground potential is the potential of the earth (in England the term is *earth*). The term *ground* is also used to mean the metal case or chassis of a piece of electronic equipment. We will discuss this in more detail later in this chapter.

The polarity of a voltage is usually discussed in reference to ground. A value above ground (for example, 10 V) is said to be +10 V, while a voltage of 10 V below ground is said to be –10 V. A battery such as shown in Figure 1-1 has a positive and a negative terminal. The positive terminal is +9 V with respect to the negative terminal side. On the other hand, the negative side is –9 V with respect to the positive side. When a battery or other DC voltage is connected in a circuit, a DC current (electrons) flows from the negative terminal of the battery through the circuit and returns to the positive terminal. This theory of current flow is called **electron flow**. Most engineers subscribe to positive current flow or conventional current flow simply because it is *conventional* (that is, it was the first theory of current flow). Energy is transferred through the circuit to a load; the results are the same regardless of the current theory used to analyze a circuit. The voltage drops have the same value, polarity, and power dissipation. In electrical circuits a battery voltage or supply voltage is denoted as **E** while voltage drops in a circuit are symbolized as **V**. Voltages can be developed from many sources, such as batteries, solar cells, thermocouples, or generators.

AC voltages (alternating current voltage) are those that vary above and below ground with respect to time. An example of an AC voltage is common house, office, and factory voltage, which changes at a rate of 60 cycles per second (Hertz). In many countries, such as Australia, the frequency is 50 Hz. Figure 1-2 illustrates a cycle of voltage from a 60-Hz source. AC voltages change with time and are also called **analog voltages.**

In this book we are interested in both analog and digital voltages. As just stated, analog voltages are those that vary with time, such as voice signals (Figure 1-3a). *Digital signals* are in

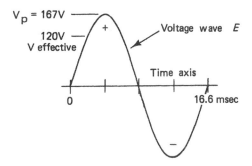

Figure 1-2 A 120-V, 60-Hz sinewave.

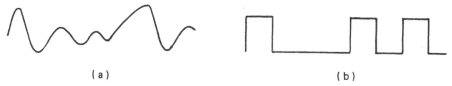

(a)   (b)

Figure 1-3 Voltages: (a) analog voltages, (b) digital voltages.

the form of pulses, called bits, that change quickly from one level of voltage to another, as shown in Figure 1-3b.

1.3 CURRENT IN AN ELECTRICAL CIRCUIT

As stated earlier, **current** or **current flow** is a movement or flow of electrons through a circuit that is caused by the electromotive force or voltage applied to the circuit. When a voltage is applied to a complete circuit, from a source, electrons move through the conductors of the circuit (energy transfer) to the load at the receiving end of the conductors. The desire is to transport the electrical signals along the conductors (wires) and have them arrive at the destination in the same configuration and with the same voltage level as those at which they left the source. In other words, the system should reproduce the input signals at the receiver and relay the input signals correctly, be they voice signals, other analog signals, DC voltages, or digital pulses.

Fiber-optic cables transport energy via photons of light energy. However, the devices that produce the signals and the devices that receive the signals depend on electrical energy for operation. The signal from the source device must be converted to light energy to be transported by the fiber cable and then converted back to electrical energy to be used by the destination device. Fiber optics will be discussed in some detail in Chapter 4.

1.4 RESISTANCE IN WIRING CIRCUITS

Resistance is the property of an electrical circuit that limits the current. Resistance can be compared to friction in a mechanical device, where the frictional drag on a body limits the speed of

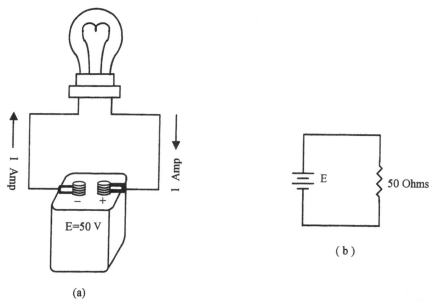

Figure 1-4 A simple electrical circuit: (a) circuit, (b) schematic diagram of the circuit.

the body and produces heat. The resistance of a material limits the number of electrons that flow in the material and the amount of energy that is transferred and causes heat in the circuit. Resistance occurs in all electrical circuits—even the best of conductors have some resistance.

The unit of resistance is the ohm and is symbolized by the Greek capital letter omega, Ω. The relationship among the voltage, current, and resistance in a circuit is called Ohm's law and is formulated as follows:

$$\text{amperes} = \frac{\text{volts}}{\text{resistance}}$$

$$I = \frac{E}{R}$$

Ohm's law states that current is directly proportional to voltage and inversely proportional to resistance. That is, increased voltage causes increased current flow and increased resistance decreases the amount of current.

Figure 1-4 depicts a simple circuit of a battery as a source and a lamp as a load. Figure 1-4a shows a pictorial of the circuit, and Figure 1-4b is a schematic diagram of the circuit. The schematic diagram of the circuit assumes that the wires to and from the 50-Ω load have no resistance.

The current is

$$I = E/R = 10 \text{ V}/50 \ \Omega = 0.2 \text{ A, or } 0.2 \text{ A}$$

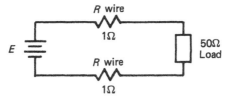

Figure 1-5 An example of an electrical circuit with wiring resistance.

The resistance is determined as follows:

$$R = E/I = 10 \text{ V}/2.0 \text{ A} = 50 \text{ }\Omega$$

We noted earlier that conductors (wires) are not perfect energy transporters but have resistance. In most cable runs wiring resistance can be ignored, but in long runs we might experience situations as depicted in Figure 1-5. The 1-Ω resistors between the source and the load represent the wiring resistance.

The current in this circuit is determined as follows. First find the total resistance:

$$R = R_1 + R_2 + R_{\text{load}} = 1 + 1 + 50 = 52 \text{ }\Omega$$

Then the current is

$$I = E/R = 10 \text{ V}/52 \text{ }\Omega = 0.0192 \text{ A}$$

The load voltage is

$$\text{Load} = I \times R_{\text{load}} = 0.192 \times 50 = 9.6 \text{ V}$$

There is a loss of 0.4 V along the line.

$$\text{Signal loss} = R_{\text{wire}} \times \text{current}$$

$$\text{SL} = 2 \text{ }\Omega \times 0.0192 \text{ A} = 0.4 \text{ V}$$

This loss of signal voltage can also be related to a loss of signal power, as we will see in Section 1.5. If the length of the wires in the preceding example were doubled, the loss would double. Obviously, very long lines would cause excessive signal loss.

The resistance of wires is determined by both length and cross-sectional area. The smaller the cross-sectional area of the conductor, the greater the resistance for a given length, and the longer the conductor, the greater the resistance for a given cross-sectional area. The properties of wire will be discussed further in Chapter 2.

1.5 POWER AND POWER LOSS

The primary purpose of transmission lines, regardless of type, is to transfer energy (power) from one device to another. **Power** is the time rate of doing work in electrical circuits. Power in watts (W) is formulated as follows:

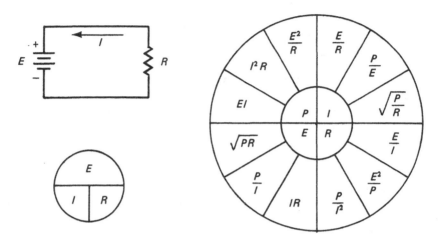

Figure 1-6   A summary of Ohm's law and power laws for the electrical properties of circuits.

$$\text{Power (watts)} = \text{current in amperes} \times \text{volts}$$

$$\text{or power} = I^2 R \text{ W}$$

Power is the unit that we use most often when relating to the levels of signal in a circuit. We usually refer to the power loss in wires as $I^2 R$ losses.

Figure 1-6 presents a summary of the formulas for both Ohm's law and the power laws.

1.6 SIGNAL-TO-NOISE RATIO

Noise is the introduction of any unwanted signal into the system. Noise may be in the background of an audio signal as an audio signal other than the desired signal. In the case of digital signals, noise may appear as analog signals or spikes that can mimic the digital pulses. Figure 1-7 illustrates the introduction of noise spikes into a digital pulse train. These noise spikes can be interpreted by a microprocessor, printer, or routing device as digital pulses and may represent a signal code other than the desired one. Although most systems contain a parity check (bit count), the introduction of two noise spikes would not necessarily be detected. In either an audio or digital system there is a threshold level below which noise can be tolerated without interfer-

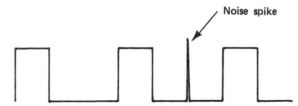

Figure 1-7   An example of noise spikes introduced into a digital pulse stream.

ence or damage to the outcome of the system. The signal-to-noise ratio (SN) in any system is formulated as

$$SN = \frac{Signal\ power}{Noise\ power}$$

For example, if an audio system had a signal level of 100 milliwatts (mW) and the noise level was 2 mW, the SN ratio would be 50:1.

1.7 INDUCTANCE AND INDUCTIVE REACTANCE IN WIRING CIRCUITS

Inductance (L) is the property of a circuit that causes an opposition to any change of current within the circuit. As electrons (current) move through a conductor, a magnetic field is produced. This magnetic field induces a voltage in the inductor, called a **counterelectromotive force** (CEMF), that opposes the current flow in the circuit. When a current tries to decrease in an inductor, CEMF is produced that tries to keep the current flowing in the circuit. CEMF can be thought of as inertia that tries to prevent change. In other words, the inductor reacts in response to a current change. Therefore, we call this phenomenon **reactance** (X).

Reactance caused by the change produces a voltage that opposes the source voltage that is producing the change. This induced voltage is formulated as

$$V_{induced} = L(di/dt)\ V$$

This formula states that the induced voltage is equal to the inductance of the coil in henries times the change in current (di) over the change in time (dt). The unit of inductance is the **henry** (H). If a current change of one ampere in 1 second produces an induced voltage of 1 volt, an inductor has an inductance of 1 henry.

Let us now put the inductive effect in context as to its effects on a digital circuit. When the source voltage (say the output of a PC) increases, the inductive reactance causes a countervoltage that slows down the voltage change at the load terminals (say a printer). Conversely, inductive reactance would slow down a decreasing voltage. The property of inductance can cause severe distortion in digital signals, where the bits must change from zero to maximum voltage and from maximum voltage to zero voltage in less than a millionth of a second. The inductive reactance of an inductor is opposition that an inductor offers to an alternating current and is formulated as

$$X_L = 2\Pi fL$$

where f represents frequency. Inspection of the formula reveals that higher frequencies result in greater values of reactance, and therefore high-frequency circuits experience more signal loss than low-frequency circuits. Wire has a small amount of inductance per meter of length; however, inductance increases as the wire length increases, much like wiring resistance. The symbol for inductance is shown in Figure 1-8.

Cross talk, the introduction of signals between conductors in close proximity to each other, is the result of the electromagnetic flux lines that are caused by the signal currents flowing

(a)   (b)

Figure 1-8 (a) Inductor, and (b) circuit symbol for an inductor.

in a conductor. Flux lines are invisible magnetic lines of force that are produced by the current (electron flow) in a circuit. The noise introduced from these flux lines, called cross talk, can cause error signals to be introduced in a data line and unwanted conversation noise in audio lines. We will discuss the methods utilized to reduce cross talk later in this chapter and in the chapters on cabling.

1.8 CAPACITANCE IN WIRING CIRCUITS

When two metals are separated by an insulator, a capacitor is formed. The symbol for a capacitor is shown in Figure 1-9. Capacitors have the ability to store energy in the form of an electrostatic charge. When one plate of the capacitor has more electrons (negative charge) than the other plate, a difference of potential exists between the plates through the insulation. The charge is the result of the force from the electron on one plate acting on the electrons on the other plate. The insulator between the plates is called a **dielectric**. The charge that is stored between the plates tends to oppose any change in circuit voltage. This opposition to a change in voltage is called **capacitive reactance**. Since a capacitor is formed by any two conductors separated by an insulator, there is capacitance between a pair of conductors in a cable, a conductor and a ground, or a conductor and a shield. The reactance of a fixed capacitor or wiring capacitance is formulated as

$$X_C = \frac{1}{2}\pi f C$$

where $\pi = 3.14$, f = frequency of the signal, and C = capacitance of the capacitor. A capacitor has a capacitance of 1 **farad** (f) when a current of 1 A causes a voltage change, across the capacitor, of 1 volt in one second.

All wires have resistance, capacitance, and inductance in varying amounts. Any or all of these factors can cause attenuation and deterioration of a pulse or an analog signal. Different

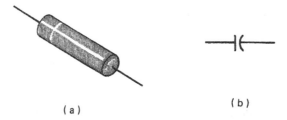

(a)   (b)

Figure 1-9 Capacitor and symbol for a capacitor.

types of copper transmission lines have different amounts of these three factors. However, it is the resistance and capacitance that result in most of the losses in transmission lines and the inductance that results in the pick-up of noise. The longer the lines, the greater the amounts of resistance, inductance, and capacitance. Increased amounts of these three factors result in increased deterioration of digital signals and increased analog signal loss.

The values of both resistance and capacitance can vary greatly based on the type of cable, wire size, shielding, and insulation. For example, cabling may vary in capacitance from a low of 8 pF per foot (25 pF per meter) to a high of 70 pF per foot (250 pF per meter). The resistance increases as the diameter of the wire decreases. The design engineer must consider these factors against economy when selecting cabling for a network.

1.9 IMPEDANCE IN WIRING CIRCUITS

The current and voltage in a resistor are always in phase with each other. That is, a maximum voltage results in a maximum current and a minimum voltage follows a minimum current. On the other hand, the current and voltage in an inductor and a capacitor are 90 degrees out of phase with each other. The current in a capacitor *leads the voltage by 90 degrees* and the current in an inductor *lags the voltage by 90 degrees*. An example of this phenomenon is depicted in Figure 1-10.

Figure 1-11 depicts a circuit with resistance, inductance, and capacitance. The voltage drops around a circuit with resistance, capacitance, and inductance are written

$$E = V + jV - jVC$$

The $+j$ and $-j$ mean +90 degrees and –90 degrees, respectively.

Impedance is the name given to the total opposition to the flow of electrical energy in a circuit and is the result of a combination of resistance, inductance, and capacitance. The symbol of impedance is Z and the unit of impedance is the ohm. Impedance is formulated as

$$Z = R + jX_L$$

The jX_L indicates that reactance must be treated different from resistance. Capacitive reactance is denoted as $-jX_C$ and inductive reactance is denoted as $+jX_L$. Again, the $+j$ and the $-j$ can be considered to indicate +90 and –90 degrees, respectively. This means that Z must be calculated as shown in Figure 1-12.

The Pythagorean theorem states that

$$Z = \sqrt{R^2 + (X_L - X_C)}$$
$$Z = \sqrt{100^2 + 100^2} = 141\ \Omega/45°$$

The 45 degrees indicates that the current and the voltage in the circuit are out of phase by 45 degrees. Impedance is rather complex in AC circuits, and it is not within the scope of this text to offer a complete study of the subject. (If you want more information, consult one of the several fine basic electronic fundamental texts available or contact the author on the Internet.) For our

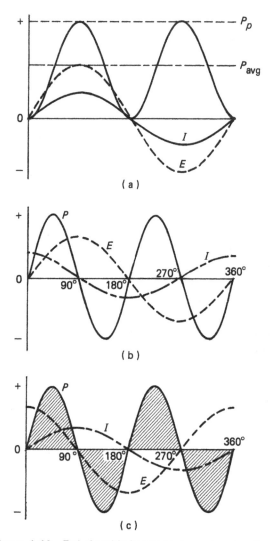

Figure 1-10 Relationship between current and voltage in an AC circuit: (a) in a resistor, (b) in a capacitor, and (c) in an inductor.

purposes we need to understand only that any output device has impedance, that all transmission lines have impedance, and that any communication device that is a load has impedance. Figure 1-13 is a summary of the reactance and impedance formulas for resistive and reactive circuits.

Transmission lines have a characteristic impedance, and loads are rated at an impedance value. We will discuss the importance of matching impedance of the transmission line to the device impedance in a later chapter.

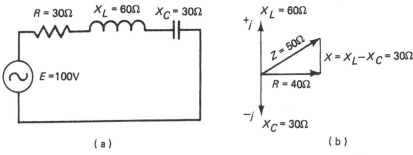

Figure 1-11   Electrical circuit with resistance, inductance, and capacitance: (a) circuit diagram, (b) vector diagram of voltages.

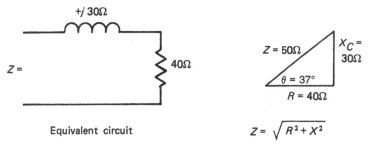

Figure 1-12   The impedance of a circuit must be calculated by using a right triangle and the Pythagorean theorem.

1.10 DIGITAL SIGNALS

Digital signals are discretely variable in the form of pulses. The pulses may represent a digital code that can be interpreted by a computer or other digital device as instructions, information, or data. Examples of instructions are add, save, or fetch. Examples of data are +3 V, –30 degrees, or $100.00. Examples of information are where to save or where to send (as an address) within the computer. Digital pulses are usually coded in a series of voltages and no voltages or ones and zeros, as shown in Figure 1-14. These digital pulses are called bits. A group of eight of these bits is called a **byte**.

The rate at which digital pulses are transmitted is called the **baud rate**, defined as bits per second or pulses per second, and is directly related to frequency. The amplitude of the pulse is the negative or positive peak voltage, as shown in Figure 1-15.

A pulse often appears to rise and fall in zero time when observed on an oscilloscope; however, this is never the case. The capacitive reactance and the inductive reactance within the circuit cause the pulse in Figure 1-14 to appear as shown in Figure 1-16. Each pulse has a rise time and a fall time as shown in Figure 1-17. The **rise time** (Figure 1-17c) is measured between the 10% point and the 90% point, and the **fall time** is measured between the 90% point and the 10%

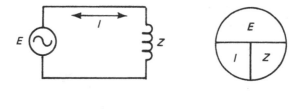

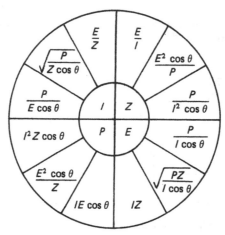

Figure 1-13   Summary of reactance and impedance formulas for reactive and resistive circuits.

Figure 1-14   A digital pulse train comprised of voltages and no voltages and representing ones and zeros.

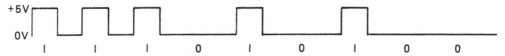

Figure 1-15   A digital pulse with a +5-V amplitude.

point. The zero and 100 percentage points are not used to measure the rise and fall times due to the difficulty of identifying their exact locations.

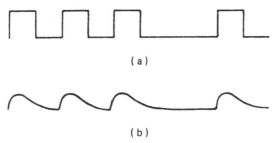

(a)

(b)

Figure 1-16 Pulse deterioration caused by the capacitance and inductance of the circuit: (a) Pulse into a transmission line, (b) distorted pulse.

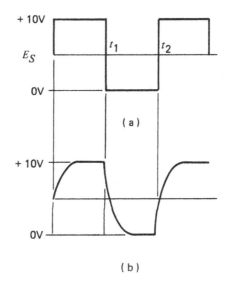

(a)

(b)

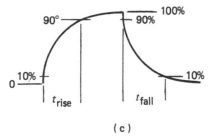

(c)

Figure 1-17 Typical digital pulse: (a) theoretical, (b) actual pulse shape, (c) rise time and fall time of a pulse.

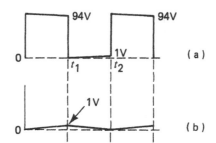

Figure 1-18 Reactance in a circuit can make the pulse unrecognizable to the destination equipment.

With very long transmission lines, where the inductance and capacitance are excessive, a pulse train may become so distorted that it becomes unrecognizable, as shown in Figure 1-18. Special equipment can sometimes reconstruct the signal. Reconstruction of digital signals is much easier than reconstruction of analog signals. For this reason analog signals are often converted to representative digital signals before transmission and reconverted to analog signals at the receiver. Circuits that perform this function are called analog to digital converters (A to D) and digital to analog converters (D to A), respectively.

1.11 ANALOG SIGNAL CONCEPTS

Analog signals are any signals other than pulses. An analog signal has a voltage that is variable with time and is usually continually variable. Some examples of analog signals are shown in Figure 1-19. An analog signal does not have to vary at a constant rate. Examples of analog signals are voice or music (audio), sampling voltages (as from a pressure gauge), or DC voltages. Signals that have a periodic repetition rate (period) have a frequency. This repetition rate or frequency, in cycles per second, is called Hertz or Hz. Frequency is formulated as

$$f = 1/T \text{ Hz}$$

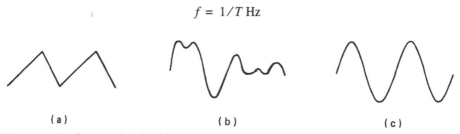

(a)   (b)   (c)

Figure 1-19 Analog signals: (a) A ramp signal, (b) a nonlinear analog signal, (c) a sinusoidal signal.

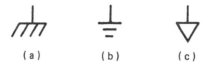

Figure 1-20 The symbols for ground:
(a) Chassis, (b) earth, (c) common.

1.12 GROUND AND GROUNDING

Ground, or *earth* as the British say, is usually referred to as the reference level for voltage levels within a system. United States government safety codes specify that all electrical equipment must be electrically connected to ground (grounded) to prevent an electrical potential between the equipment and ground and between pieces of equipment. Equipment grounding is a safety precaution designed to protect both people and equipment. The symbol for ground is shown in Figure 1-20. The chassis of most equipment is grounded, and the return path for current flow is in the chassis. This reduces the need for returned wires from the components. Most voltages in electronic equipment are measured to ground (the chassis).

Grounding of electrical equipment is usually accomplished through the power plug. For 120-V connections the center prong of the three-prong plug is ground (Figure 1-21). The insulation color of the ground wire in a power lead to equipment is usually green or green and yellow. Grounding of the shielding wire in telecommunication cable is important to assure the transmission of electrical signals along a cable without interference from the electromagnetic radiation from other transmission lines and electrical equipment. This interference, called *cross talk*, can originate from adjacent lines, electrical motors, PCs, fluorescent lights, etc. The term cross talk

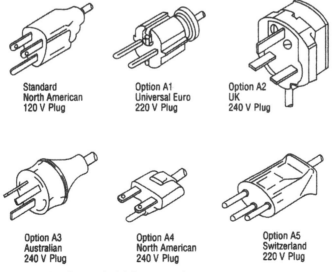

Standard
North American
120 V Plug

Option A1
Universal Euro
220 V Plug

Option A2
UK
240 V Plug

Option A3
Australian
240 V Plug

Option A4
North American
240 V Plug

Option A5
Switzerland
220 V Plug

Figure 1-21 Grounded AC power plugs.

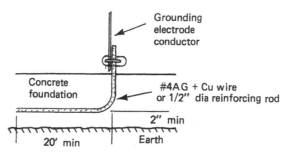

Figure 1-22 A plant grounding system.

originated from the phenomenon of the conversation of an adjacent line being audible on the other.

While equipment grounding is primarily for the protection of people from electrical shock, there are other compelling reasons for grounding. Grounding provides a low-impedance path for electrical energy. In summary, grounding provides the following:

1. Protection of people from electrical shock in the event of an internal short in equipment
2. Protection of semiconductor devices from excessive static voltage buildup
3. A safe path for electrical energy from lightning to protect both equipment and people
4. A low-resistance path around the signal-carrying wires for low-frequency electromagnetic energy from sources such as power lines, lights, and motors
5. A low-resistance path around the signal-carrying wires for electromagnetic radiation from high-frequency electromagnetic waves from computers, other transmission lines, radiated signals (radio, TV), etc.

The grounding system for a facility should maintain all the grounds of all telecommunication equipment, other electronic equipment, all electrical equipment, and all electrical power at the same potential, within the closely prescribed limits of the National Electrical Code (NEC).

The **earth ground system** is the reference for all grounds within a building. The earth ground is established by inserting bronze rods into the earth or bronze or copper wire into the concrete foundation of a building. This part of the ground system is the most difficult to establish to assure long-range effectiveness because of the wide range of soil types and the varying moisture content of soil. The moisture content of soil and the minerals within the soil determine the resistance of the soil, and thereby the effectiveness of the ground in maintaining a low-voltage interface. Figure 1-22 depicts an example of a plant grounding system referenced from the NEC for proper grounding. Whenever possible, the connection to the ground electrode should be less than 1 foot below the surface of the soil, and the grounding electrode should extend at least 10 feet below ground. Ground in soil types other than moist clay requires special installation techniques. For example, grounding in shallow soil requires that grounding cable be laid in trenches and the soil compacted. Grounding systems must be a primary consideration

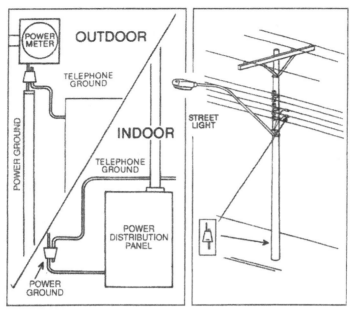

Figure 1-23   Power meter, distribution panel, and street light grounding examples. (*Source:* Courtesy AMP Inc.)

when designing a new facility. Figure 1-23 illustrates examples of telephone grounding, street light grounding, and service entrance grounding. Ground conductors must be electrically connected to prevent any resistance between the conductors. Connectors must withstand physical strains and weather variations without reduction of conductivity. When aluminum and copper conductors are bonded, the connector must keep the two conductors physically isolated to prevent battery action while maintaining high conductivity. When two dissimilar metals, such as copper and aluminum, are bonded, a virtual battery is formed, producing corrosion, which causes increased resistance and decreased conductivity between the metals. Figure 1-24 depicts three examples of ground connections.

In newer construction, architects often require a grid type of grounding comprised of heavy copper or aluminum conductors or metal rods. Figure 1-25 illustrates methods of bonding subgrounding conductors to the grid.

All grounding installations must be in compliance with the latest edition of the NEC handbook.

1.13 CROSS TALK IN WIRING

Electromagnetic pick-up and radiation can cause serious problems in telecommunication systems, such as signal distortion, noise in conversations, and breach of security from a system. Some types of cables are protected from inductively induced signals (cross talk) from adjacent lines. For example, pairs of wires are usually twisted to reduce inductive effects, and cables can

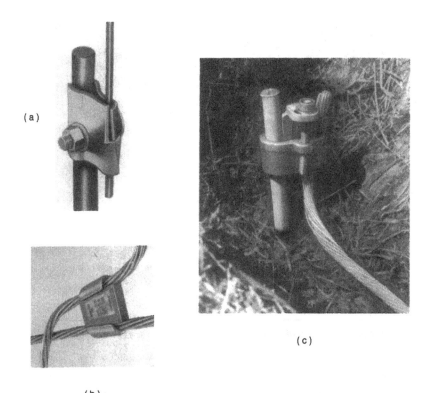

(a)

(c)

(b)

Figure 1-24 Grounding connectors: (a) rod to small conductor, (b) conductor to conductor, (c) rod to primary ground cable. (*Source:* Courtesy AMP Inc.)

be shielded from outside electromagnetic lines by surrounding the signal-carrying wires with a braided or solid metal shield. There are also installation procedures that can reduce the effects of magnetic induction noise, such as

1. Shielding the cables
2. Never running data cable in a conduit with power cables
3. Using proper grounding of equipment and cables to protect against lighting and surge voltages and to provide shielding against outside signals

Grounding will be discussed in greater detail in Chapters 2 through 4.

1.14 ATTENUATION OF SIGNAL INFORMATION

As stated earlier, wire has resistance, inductance, and capacitance. All these factors attenuate both digital and analog signals. The attenuation can be measured in either a reduction of voltage or a loss of power. This loss is usually referred to as a decibel loss. The Bel is the logarithm of the power input to the power output of a system (in the case of a cable, the power into the cable

Typical Grid Connections

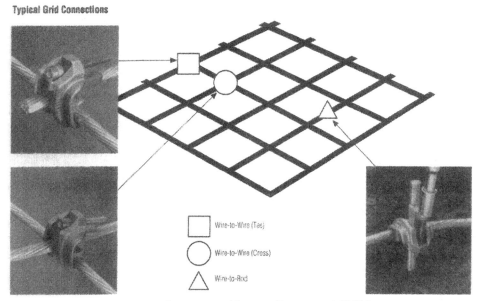

Figure 1-25 Grid-type grounding system (*Source:* Courtesy of AMP Inc.

at the source and power output at the receiver). The Bel unit is such a huge number that the decibel (1/10 Bel) is usually used for calculations.

The formula for the decibel (dB) is

$$dB = 10\log(P_{out}/P_{in})$$

The decibel gain or loss can also be formulated using voltages:

$$dB = 20\log V_{out}/V_{in}$$

For example, if a signal of 1 W was put into a line with a resulting attenuation to 0.5 W at the receiver end, the loss in decibels is

$$dB = 10\log 0.5/1 = 10\log 0.5 = -3$$

We would say that the signal had a −3-dB loss.

As another example, suppose that a 1-V signal were injected into a transmission line with a reduction of to 0.707 V at the receiver end. The dB gain is

$$dB = 20\log 0.707/1 = 20\log 0.707 = -3$$

Again we would say that the attenuation within the line was −3 dB. We might note that −3 dB is also a 50% power loss. Figure 1-26 is an example of decibel gains and losses in a circuit.

Communication systems often have to be designed to accommodate a combination of analog and digital signals. The designer and cabling technician are often required to deal with

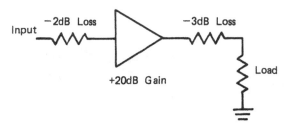

Figure 1-26   dB gains and losses in a circuit.

twisted pairs, coaxial cables, and/or fiber-optic cables. The techniques to perform installation and maintenance of these designs are discussed in later chapters.

1.15 INSULATION OF CONDUCTORS

Insulation is the nonconductive material that encases a wire or cable. Insulation materials are comprised of compounds that have properties that are Underwriters Laboratory (UL) rated and Canadian Standards Association (CSA) approved to prevent certain environment hazards while satisfying special electrical requirements. The electrical requirements might be fire resistance, weather resistance, pressure resistance, etc. The environmental hazards might be that the insulator gives off toxic gases in a fire. Wires are individually insulated for ratings such as minimum breakdown voltage, wiring capacitance, and maximum temperature. The primary purpose of any insulation is to prevent the short circuiting of wires to other wires or ground, which could cause signal loss, damage to equipment, and possibly fire. When more than one conductor is bundled into a cable, the insulating material for the cable is called a jacket. The purpose of the jacket, other than holding the wires together, is to protect the cable.

Wire and cable insulating coverings are made from several insulating materials and compounds. Insulating coverings are rated by the Underwriters Laboratory, a private rating company that is the industry standard for rating of consumer products for properties such as electrical characteristics, heat resistance, chemical characteristics, reliability, and safety. The following are the most common insulating materials used in the insulation of cables and their properties.

- *Vinyl:* Vinyl is sometimes referred to as PVC or polyvinyl chloride. Certain formulas have temperature ratings from –40°C to +105°C. Other common vinyls may have ratings from –20°C to +60°C. There are many formulations for different applications. The formulation affects both the electrical properties and the pliability, which can vary from rock hard to puttylike.
- *Polyethylene:* This material has excellent electrical properties, with a low dielectric value (low capacitance). Flexibility can vary from soft to rock hard. This insulation has excellent moisture resistance and can be formulated to withstand extreme weather conditions.
- *Teflon:* This material has excellent electrical properties, temperature range, and chemical resistance. The material is not suited for high-voltage applications or for an environment

within nuclear radiation. The cost of Teflon insulation is approximately 10 times that for comparable vinyl insulation.

• *Polypropylene:* This insulation is similar to polyethylene in electrical properties but is typically harder than polyethylene. It is suitable for thin-wall insulation. Most UL ratings call for 60°C.

• *Silicone:* This is a very soft insulation with a temperature range of –80°C to +90°C. It has excellent electrical properties along with ozone resistance, low moisture absorption, weather resistance, and radiation resistance. However, it has low mechanical strength and poor scuff resistance and costs from $5.00 to $8.00 per pound compared with $1.00 per pound for other insulation.

• *Neoprene:* The maximum temperature range of this material can vary from 55°C to +90°C. The electrical properties are not as good as other insulating materials, resulting in the need for thicker insulation. Typically this material is used as a coating for separate lead wires or cable jackets.

• *Rubber:* Both natural rubber and synthetic rubber compounds can be used for insulation and cable jackets. The material is formulated for many different applications and many different temperature ranges.

Table 1-1 presents the properties of rubber insulation. Table 1-2 summarizes the properties of plastic insulation. Table 1-3 gives the nominal temperature range of various insulating materials when used as wire insulation or cable jackets.

Table 1-1   Comparative properties of rubber insulation

	Rubber	Neoprene	Hypalon (Chloro-sulfonated Polyethylene)	EPDM (Ethylene Propylene Diene Monomer)	Silicone
Oxidation resistance	F	G	E	G	E
Heat resistance	F	G	E	E	O
Oil resistance	P	G	G	F	FG
Low temperature flexibility	G	F,G	F	G,E	O
Weather, sun resistance	F	G	E	E	O
Ozone resistance	P	G	E	E	O
Abrasion resistance	E	G,E	G	G	P

Table 1-1   Comparative properties of rubber insulation (Continued)

	Rubber	Neoprene	Hypalon (Chloro-sulfonated Polyethylene)	EPDM (Ethylene Propylene Diene Monomer)	Silicone
Electrical properties	E	P	G	E	O
Flame resistance	P	G	G	P	F,G
Nuclear radiation resistance	F	F,G	G	G	E
Water resistance	G	E	G,E	G,E	G,E
Acid resistance	F,G	G	E	G,E	F,G
Alkali resistance	F,G	G	E	G,E	F,G
Gasoline, kerosene, etc. (aliphatic hydrocarbons) resistance	P	G	F	P	P,F
Benzol, Tuluol, etc. (aromatic hydro-carbons) resistance	P	P,F	F	F	P
Degreaser solvents (halogenated hydro-carbons) resistance	P	P	P,F	P	P,G
Alcohol resistance	G	F	G	P	G

P = poor F = fair G = good E = excellent O = outstanding

These ratings are based on average performance of general purpose compounds. Any given property can usually be improved by the use of selective compounding.

Source: Courtesy Belden Corporation.

Table 1-2 Comparative properties of plastic insulation

	PVC	Low-Density Poly-ethylene	Cellular Poly-ethylene	High-Density Polyethylene	Poly-propylene	Poly-urethane	Nylon	Teflon
Oxidation resistance	E	E	E	E	E	E	E	O
Heat resistance	G,E	G	G	G	E	E	E	O
Oil resistance	F	G	G	G,E	F	E	E	O
Low temperature flexibility	P,G	G,E	E	E	P	G	G	O
Weather, sun resistance	G,E	E	E	E	E	G	E	O
Ozone resistance	E	E	E	E	E	E	E	E
Abrasion resistance	F,G	F,G	F	E	F,G	O	E	E
Electrical properties	F,G	E	E	E	E	P	P	E
Flame resistance	E	P	P	P	P	P	P	O
Nuclear radiation resistance	G	G	G	G	F	G	F,G	P
Water resistance	E	E	E	E	E	P,G	P,F	E
Acid resistance	G,E	G,E	G,E	G,E	E	F	P,F	E
Alkali resistance	G,E	G,E	G,E	G,E	E	F	P,F	E
Gasoline, kerosene, etc. (aliphatic hydrocarbons) resistance	P	P,F	P,F	P,F	P,F	G	G	E
Benzol, Tuluol, etc. (aromatic hydrocarbons) resistance	P,F	P	P	P	P,F	P	G	E
Degreaser solvents (halogenated hydrocarbons) resistance	P,F	P	P	P	P	P	G	E
Alcohol resistance	G,E	E	E	E	E	P	P	E

P = poor F = fair G = good E = excellent O = outstanding

These ratings are based on average performance of general purpose compounds. Any given property can usually be improved by the use of selective compounding.

Source: Courtesy Belden Electronics Division.

Table 1-3   Nominal temperature range for insulating and jacketing compounds

Compound	Normal Low	Normal High	Special Low	Special High
Chlorosulfonated polyethylene	–20°C	90°C	–40°C	105°C
EPDM (ethylene propylene rubber)	–55°C	105°C	—	—
Neoprene	–20°C	60°C	–55°C	90°C
Polyethylene	–60°C	80°C	—	—
Polypropylene	–40°C	105°C	—	—
Rubber	–30°C	60°C	–55°C	75°C
Teflon	–70°C	200°C	—	260°C
Vinyl	–20°C	80°C	–55°C	105°C
Silicone	–80°C	150°C	—	200°C
Halar	–70°C	150°C	—	—

Source: Courtesy Belden Electronics Division.

1.16 SUMMARY

The proper planning and installation of telecommunication wiring is a complex task and should not to be attempted without the skills and knowledge necessary to complete the task successfully. The material in the following chapters, if studied, will greatly improve your chance of a successful installation.

QUESTIONS

1. What unit is used most often for cable signal loss?
2. What causes cross talk?
3. Why do wires have capacitance?
4. List several factors that would be considered in the selection of wiring insulation.
5. Define the term *analog* and give an example of an analog signal.
6. Define the term *bit*.
7. Define the term *digital signal* and give an example.

CHAPTER 2

Transmission Media: Twisted Pair

2.1 INTRODUCTION

As a variety of data processing equipment is added to a company's inventory, management recognizes the need to tie the units together for better utilization and economy. This leads most companies to develop the equipment into some form of **centralized network**. When individual equipment such as host computers, PCs, workstations, scanners, plotters, and printers are connected into a common data-sharing system, it is called a **local area network** (LAN). The connection of a LAN requires specialized wiring to assure the transmission of telecommunication data between the various equipment and the proper operation of each unit.

The wiring used may be residential telephone cabling, new high-frequency twisted-pair cabling, coaxial cabling, fiber-optic cabling, wireless transmission, or a combination of these. We will examine the characteristics of cabling in the following chapters.

This chapter focuses on copper wire twisted-pair-type cabling. Chapter 3 is devoted to coaxial cabling, and Chapter 4 is devoted to fiber-optic cabling. A summary of the three types of wiring media is given at the end of Chapter 4.

2.2 UNSHIELDED TWISTED-PAIR (UTP) CABLING*

Twisted pair is the cabling medium most often used in the telephone and telecommunication industry. It is comprised of two insulated copper wires twisted approximately 20 turns per foot (Figure 2-1). Most telephone cable is comprised of solid #24-gauge copper wires, each insulated and twisted in a pair. The most common cabling has two pairs of wires enclosed in a tough outer

---

* In this and the following chapter the reader will encounter numerous new terms and abbreviations of terms. These are explained in detail in the Glossary.

Figure 2-1 A twisted-wire cable.

jacket. Each of the four wires is color coded, usually red, green, black, and yellow. Only two of the wires are needed for telephone function, leaving the other two available for a second line or other functions.

Multiple-pair cabling uses wiring with two colors. Each wire in a multiple-pair cable will have one of the basic colors with one of many colored stripes. Unfortunately for the wiring professional, there is no one color standard for color coding of wiring. Tables 2-1 through 2-6 represent the standards. Table 2-1 is Belden's standard for single conductors in a cable. Table 2-2 is the color code of the Insulated Cable Engineers Association (ICEA). Table 2-3 is the coding for Belden's paired cable. Table 2-4 is the standard coding of Western Electric Company. Table 2-5 is the modified Western Electric coding for Western Electric's Data Twist® cabling. Table 2-6 is the standard coding for IBM RISC System/6000 cables. It is important that the technician utilize the correct coding when installing or testing cabling.

Table 2-1 Colors for single conductors in a 12-conductor cable

Color Code Chart No. 1

Cond.	Color
1st	Black
2nd	White
3rd	Red
4th	Green
5th	Brown
6th	Blue
7th	Orange
8th	Yellow
9th	Purple
10th	Gray
11th	Pink
12th	Tan

18 Gage conductors in cables 8446 through 8449 are Black and White.

Source: Courtesy Belden Inc., Electronics Division

Table 2-2   A sample color chart for ICEA standards

Color Code Chart No. 2 and 2R (ICEA—Insulated Cable Engineers Association Standard) 2R—These cables feature a ring band striping. 2—These cables feature a spiral stripe.

Cond.	Color	Cond.	Color	Cond.	Color	Cond.	Color
1st	Black	14th	Green/White	27th	Blue/Black/White	40th	Red/White/Green
2nd	White	15th	Blue/White	28th	Black/Red/Green	41st	Green/White/Blue
3rd	Red	16th	Black/Red	29th	White/Red/Green	42nd	Orange/Red/Green
4th	Green	17th	White/Red	30th	Red/Black/Green	43rd	Blue/Red/Green
5th	Orange	18th	Orange/Red	31st	Green/Black/Orange	44th	Black/White/Blue
6th	Blue	19th	Blue/Red	32nd	Orange/Black/Green	45th	White/Black/Blue
7th	White/Black	20th	Red/Green	33rd	Blue/White/Orange	46th	Red/White/Blue
8th	Red/Black	21st	Orange/Green	34th	Black/White/Orange	47th	Green/Orange/Red
9th	Green/Black	22nd	Black/White/Red	35th	White/Red/Orange	48th	Orange/Red/Blue
10th	Orange/Black	23rd	White/Black/Red	36th	Orange/White/Blue	49th	Blue/Orange/Red
11th	Blue/Black	24th	Red/Black/White	37th	White/Red/Blue	50th	Black/Orange/Red
12th	Black/White	25th	Green/Black/White	38th	Black/White/Green		
13th	Red/White	26th	Orange/Black/White	39th	White/Black/Green		

Source: Courtesy Belden Inc., Electronics Division

Table 2-3   A color chart for paired cable

Color Code Chart No. 3 for Paired Cables (Belden Standard)

Pair No.	Color Combination	Pair No.	Color Combination	Pair No.	Color Combination	Pair No.	Color Combination
1	Black & Red	11	Red & Yellow	21	White & Brown	31	Purple & White
2	Black & White	12	Red & Brown	22	White & Orange	32	Purple & Dark Green
3	Black & Green	13	Red & Orange	23	Blue & Yellow	33	Purple & Light Blue
4	Black & Blue	14	Green & White	24	Blue & Brown	34	Purple & Yellow
5	Black & Yellow	15	Green & Blue	25	Blue & Orange	35	Purple & Brown
6	Black & Orange	16	Green & Yellow	26	Brown & Yellow	36	Purple & Black
7	Black & Orange	17	Green & Brown	27	Brown & Orange	37	Gray & White
8	Red & White	18	Green & Orange	28	Orange & Yellow		
9	Red & Green	19	White & Blue	29	Purple & Orange		
10	Red & Blue	20	White & Yellow	30	Purple & Red		

Source: Courtesy Belden Inc., Electronics Division

Table 2-4   A color chart for Western Electric standards

Color Code Chart No. 5 for Paired Cables (Western Electric Standard)

Pair No.	Color Combination	Pair No.	Color Combination	Pair No.	Color Combination	Pair No.	Color Combination	Pair No.	Color Combination
1	White / Blue Stripe Blue / White Stripe	6	Red / Blue Stripe Blue / Red Stripe	11	Black / Blue Stripe Blue / Black Stripe	16	Yellow / Blue Stripe Blue / Yellow Stripe	21	Purple / Blue Stripe Blue / Purple Stripe
2	White / Orange Stripe Orange / White Stripe	7	Red / Orange Stripe Orange / Red Stripe	12	Black / Orange Stripe Orange / Black Stripe	17	Yellow / Orange Stripe Orange / Yellow Stripe	22	Purple / Orange Stripe Orange / Purple Stripe
3	White / Green Stripe Green / White Stripe	8	Red / Green Stripe Green / Red Stripe	13	Black / Green Stripe Green / Black Stripe	18	Yellow / Green Stripe Green / Yellow Stripe	23	Purple / Green Stripe Green / Purple Stripe
4	White / Brown Stripe Brown / White Stripe	9	Red / Brown Stripe Brown / Red Stripe	14	Black / Brown Stripe Brown / Black Stripe	19	Yellow / Brown Stripe Brown / Yellow Stripe	24	Purple / Brown Stripe Brown / Purple Stripe
5	White / Gray Stripe Gray / White Stripe	10	Red / Gray Stripe Gray / Red Stripe	15	Black / Gray Stripe Gray / Black Stripe	20	Yellow / Gray Stripe Gray / Yellow Stripe	25	Purple / Gray Stripe Gray / Purple Stripe

Source: Courtesy Belden Inc., Electronics Division

Table 2-5 A color chart for Western Electric Twist®-type cable

Color Code Chart No. 6

Position No.	Color	Position No.	Color
1	Brown	13	White/Orange
2	Red	14	White/Yellow
3	Orange	15	White/Green
4	Yellow	16	White/Blue
5	Green	17	White/Purple
6	Blue	18	White/Gray
7	Purple	19	White/Black/Brown
8	Gray	20	White/Black/Red
9	White	21	White/Black/Orange
10	White/Black	22	White/Black/Yellow
11	White/Brown	23	White/Black/Green
12	Drain	24	White/Black/Blue

IBM RISC System/6000 Cables Color Code

Cond.	Color	Pair No.	Color Combination
1st 2nd	White over Blue White over Orange	1	White over Blue Blue over White
3rd 4th	White over Green White over Brown	2	White over Orange Orange over White
5th 6th	White over Gray White over Red	3	White over Green Green over White
7th	White over Yellow		

Source: Courtesy Belden Inc., Electronics Division

Table 2-6 IBM RISC system 6000 cable color code

Data Twist™ Color Code Chart (Modified Western Electric)

Pair No.	Color Combination	Pair No.	Color Combination	Pair No.	Color Combination	Pair No.	Color Combination	Pair No.	Color Combination
1	White/Blue Stripe Blue	6	Red/Blue Stripe Blue/Red Stripe	11	Black/Blue Stripe Blue/Black Stripe	16	Yellow/Blue Stripe Blue/Yellow Stripe	21	Purple/Blue Stripe Blue/Purple Stripe
2	White/Orange Stripe Orange	7	Red/Orange Stripe Orange/Red Stripe	12	Black/Orange Stripe Orange/Black Stripe	17	Yellow/Orange Stripe Orange/Yellow Stripe	22	Purple/Orange Stripe Orange/Purple Stripe
3	White/Green Stripe Green	8	Red/Green Stripe Green/Red Stripe	13	Black/Green Stripe Green/Black Stripe	18	Yellow/Green Stripe Green/Yellow Stripe	23	Purple /Green Stripe Green/Purple Stripe
4	White/Brown Stripe Brown	9	Red/Brown Stripe Brown/Red Stripe	14	Black/Brown Stripe Brown/Black Stripe	19	Yellow/Brown Stripe Brown/Yellow Stripe	24	Purple /Brown Stripe Brown/Purple Stripe
5	White/Gray Stripe Gray/White Stripe	10	Red/Gray Stripe Gray/Red Stripe	15	Black/Gray Stripe Gray/Black Stripe	20	Yellow/Gray Stripe Gray/Yellow Stripe	25	Purple /GrayStripe Gray/Purple Stripe

Source: Courtesy Belden Inc., Electronics Division

Twisted-pair wiring specifically designated for data transmission is usually between 22 and 26 gauge, according to the American Wire Gauge (AWG) measurement system. The AWG number is the number that represents the wire diameter. The AWG number is used in Table 2-7 to give a host of information about each wire, such as the AWG number of the wire, the diameter of the wire, the resistance, and the weight of the wire. Notice that the larger the wire gauge number, the smaller the diameter of the wire and the higher the resistance of the wire. For example, 22-gauge wire has a diameter of 25.36 mil (25.36 1/1000 in) or 0.02535 inches and #24 wire has a diameter of 20.1 mil or 0.0201 in. The resistance of #22 wire is 16.14 Ω per 1000 ft and the resistance of #24 wire is 25.67 Ω per 1000 ft. Of course, long runs of #22 wire would be considerably heavier than #24 wire. On the other hand, because of the higher resistance of the #24 wire the signal would be attenuated considerably more over a #24 wire as compared to the same run of #22 wire. When in-place telephone cabling is used for data cabling, the installed #24 wire will suffice for most applications. However, given a choice, the cable installer should use #22 wire to reduce the effects of wiring resistance on signal transmission.

Table 2-7   A partial table of the American Wire Gauge showing the characteristics of wire

Gage No.	Diameter (mils)	Cross Section Circular (mils)	Cross Section Square inches	Ohms per 1000 Feet 25°C (=77°F)	Ohms per 1000 Feet 65°C (=149°F)	Pounds per 1000 Feet	Current Carrying Capacity @ 700 CM per Amp	Ohms per 1000 Ft @ 20°C
18	40.0	1620.0	0.00128	6.51	7.51	4.92	2.93	5.064
19	36.0	1290.0	0.00101	8.21	9.48	3.90	2.32	6.385
20	32.0	1020.0	0.000802	10.4	11.9	3.09	1.84	8.051
21	28.5	810.0	0.000636	13.1	15.1	2.45	1.46	10.15
22	25.3	642.0	0.000505	16.5	19.0	1.94	1.16	12.80
23	22.6	509.0	0.000400	20.8	24.0	1.54	0.918	16.14
24	20.1	404.0	0.000317	26.2	30.2	1.22	0.728	20.36
25	17.9	320.0	0.000252	33.0	38.1	0.970	0.577	25.67
26	15.9	254.0	0.000200	41.6	48.0	0.769	0.458	32.37
27	14.2	202.0	0.000158	52.5	60.6	0.610	0.363	40.81
28	12.6	160.0	0.000126	66.2	76.4	0.484	0.288	51.47

Older telephone cabling is limited to approximately 24,000 kilobits per second (kbps). New coding techniques, such as *high bit-rate digital subscriber line* (HDSL), allow transmission of T1 services of 1.544 Mbps over a carrier service area.

T1 is a TDM digital channel carrier that operates at the rate of 1.544 Mbps over two twisted-pair wiring. T1 is often confused with DS-1, the specifications for framing and transmitting a 1.544-Mbps bit stream consisting of twenty-four 64-kbps DS-0 channels.

DS-1 was originally specified to be transmitted over T1 wiring but has subsequently been specified on other types of media as well. As a result, the terms T1 and DS-1 are often used interchangeably. The proposed G.shdsl is a multirate symmetric DSL technology that will allow the transmission of 192 to 2.304 kbps up to a distance of 24,000 feet (7315 meters). These and other pulse modulated technologies have extended the bandwidth over copper pairs to allow the simultaneous transmission of data, voice, and video over one or two copper pairs. We will discuss DSL in detail in Chapter 7.

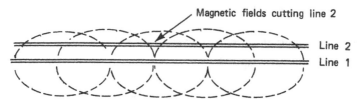

Figure 2-2   An illustration of how cross talk occurs between adjacent lines.

2.3   CROSS TALK ON TWISTED-PAIR CABLE

Twisted-pair cable, like any other copper cable, generates electromagnetic energy as signals pass through it, and picks up signals called cross talk from electromagnetic waves generated from other sources. As we stated in Chapter 1, electromagnetic pick-up can be from any electrical device, such as audio or data lines, motors, fluorescent light, and computers, to name a few. The term cross talk, as you may remember from Chapter 1, originated from the phenomenon of the conversation from adjacent telephone wires being slightly audible on another line. However, the term has been expanded to mean any type of electromagnetic interference that can cause signal distortion. While electromagnetic pick-up is usually only an annoyance in telephone conversations, it can be devastating to digital data signals. The noise spikes caused by external electrical devices can mask some of the bits or appear as additional bits that can confuse the data coding of the computer. Figure 2-2 illustrates how the signals from one line can magnetically couple to another.

There is more than one definition for cross talk. **Near-end cross talk** (NEXT) is interference where the interfering signal is traveling in the opposite direction to the desired signal. **Far-end cross talk** (FEXT) is the coupling of received signals on the received signals of an adjacent pair.

Twisted-pair cable can be shielded to reduce electromagnetic pick-up and radiation by use of metallic braid, as shown in Figure 2-3.

2.4   SHIELDING OF TRANSMISSION LINES

Shielding is an important part of the installation of any cabling and aids in producing a low signal-to-noise ratio in the equipment. The signal-to-noise ratio is the amount of signal voltage or power in ratio to the amount of noise voltage or power. When new equipment is installed or cables or equipment are moved, the grounds to the cables and the equipment must be tested.

The metallic shield of a cable must be grounded for the shielding to be effective in preventing electromagnetic energy from being emitted from the cable or picked up from outside sources. Shielding is most effective when each pair of wires is shielded. Overall shielding of a multiple conductor cable is only effective in keeping internally produced electromagnetic radiation within the cable and preventing interference from outside the cable. It does nothing to prevent cross talk between wires within the cable.

Figure 2-3   Shielded twisted-pair cable.

To be most effective, shielding for telecommunication cables should be grounded at the source only. It would seem that if shielding is effective in limiting electromagnetic radiation from data communication wires and pick-up from other sources, more shielding would be better. However, grounding of the shield at both ends creates an electrical path through the ground between the source and the destination equipment, as depicted in Figure 2-4. The electrical path between equipment is called a *ground loop* and results in a difference of potential (voltage) between the source and the destination equipment. This voltage, called the *ground loop potential*, can be as much as several volts. Ground loop voltage causes a current to flow in

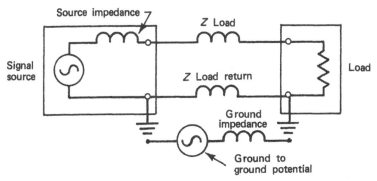

Figure 2-4   Shielding transmission line at both ends creates a ground loop.

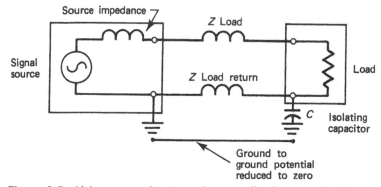

Figure 2-5 Using a capacitor to produce an effective ground at both the source and the destination.

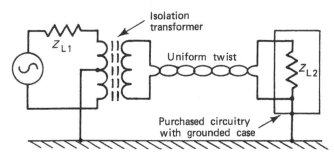

Figure 2-6 The use of an isolation transformer to prevent a ground loop path.

the low resistance of the shield, resulting in electromagnetic waves that are picked up as noise by the transmission wires.

Although it is desirable to shield cables only at the source, certain equipment devices must be grounded to operate correctly. One way to satisfy this need and the resulting grounding at both ends of the cable is to place a capacitor between the earth ground connection and the equipment ground (Figure 2-5). The capacitor blocks the DC voltage from causing a ground loop voltage while passing the varying electromagnetic signals to ground. Another possible solution to the multiple-ground problem is to use an isolation amplifier, as shown in Figure 2-6. The isolation amplifier has no physical connection between the input and the output and has no return ground loop path between the devices. Good-quality shielded-pair cable is expensive. However, the expense is sometimes justified as such cable must meet rigid manufacturing specifications with regard to certain features:

1. the diameter and the strength of each conductor
2. the properties of the insulation
3. the twisting of each pair

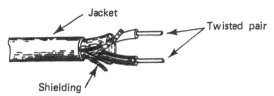

Figure 2-7 An example of IBM twinaxial shielded cable.

4. the use of a metallic foil to shield each pair
5. the use of an outer layer of insulation to shield and seal the entire cable

An example of quality shielding is the IBM twinaxial cabling for IBM cabling systems. Twinaxial cable is comprised of four wires—a red and green twisted pair and a black and orange twisted pair—encased in a braided shield and covered by an outer insulation jacket (Figure 2-7).

In summary, shielded cable is subject to less cross-talk interference and radiates less electromagnetic signal. It is therefore less subject to loss of communication data that would be caused by stray pick-up from other cables. Shielded cables also offer greater information security since they radiate less electromagnetic signal to other cables.

Shielding is not perfect, and the best of efforts will still result in some level of noise on a transmission line. The amplitude of the noise that occurs on a line with respect to the signal information is called the signal-to-noise ratio. Both analog and data equipment are rated as to the level of signal-to-noise that can be tolerated without sacrificing equipment performance. Either shielded or unshielded twisted pair can be bundled into cables with hundreds of wire pairs, as shown in Figure 2-8.

2.5 APPLICATIONS AND FUNCTIONS OF TWISTED PAIRS

As noted earlier, twisted-pair cable has been used for voice transmission by telephone for years. Each pair of wires can carry more than one voice channel. This is accomplished by placing (modulating) the voice frequencies (20 to 3000 Hz) on a higher carrier frequency. A carrier frequency is a higher frequency on which the intelligence or information is induced. Each carrier has a different frequency, with an allotted voice bandwidth. The audio signals are modulated onto the carrier, resulting in an envelope of frequencies. Although the different audio signals are

Figure 2-8 An illustration of a cable with many shielded pairs.

within the same frequency range, they do not interfere with each other as the act of modulation places them within the frequency range of their carrier.

Modulation is the process of modifying a carrier frequency in rhythm to the audio frequency. The modification of the carrier (modulation) can be by a change of amplitude, a change of frequency, or a change of phase. The modulation of a carrier by another signal produces other frequencies. For example, a carrier frequency of 1 MHz modulated by a 1-kHz audio signal would result in the following frequencies:

$$F_{carrier}, F_{audio}, F_{carrier} + F_{audio}, F_{carrier} - F_{audio}$$

$$1\,MHz, 1\,kHz, 1.001\,MHz, \text{ and } 999 \text{ kHz}$$

These frequencies make up the modulation envelope, and the bandwidth of the modulation would be 999 kHz to 1.001 MHz or 2000 Hz around the 1-MHz carrier.

The modulation of any two frequencies (F_1 and F_2) results in the development of the two new frequencies: the sum and the difference of the carrier and the modulating frequency. Figure 2-9 illustrates the three methods of modulation. Figure 2-9a denotes *amplitude modulation*, where the amplitude of the carrier is varied by the modulation. Figure 2-9b illustrates *frequency modulation*, where the frequency of the carrier is varied in rhythm with the modulating frequency. Figure 2-9c illustrates the effects of *phase modulation*, where the phase of the carrier is shifted in rhythm with the modulation. Phase modulation changes the time relation of the carrier. When the modulating signals are pulses, the carrier is turned on and off (pulse modulation).

Modulation makes each frequency channel unique, much like radio stations or TV channels. This type of channeling is called *frequency division multiplexing* (FDM). The Federal Communications Commission (FCC) designates the frequency channels available for multiplexing in communications.

One wire pair can carry many voice channels simultaneously using high-frequency carrier signals. A disadvantage is that the higher the frequency of the carrier, the greater the amount of signal attenuation. This results in the need for repeaters along long lines to amplify the signal. Utilizing pulse-coded audio-modulated techniques, twisted pair can be used for limited distances, for xDSL applications to transmit a number of audio channels, for digital data, and for video signals.

When several signals are transmitted over the same media but at different times, the method is called *time division multiplexing* (TDM). With TDM the signals are transmitted at different time frames that are synchronized between the transmitter and the receiver. For example, three terminals that require access to the same host computer appear to each be in direct connection with the computer. TDM does not change the frequency of the signals and does not create any new signals. Figure 2-10 illustrates time division multiplexing. Devices that control multiplexing at each end of a telecommunication system are called multiplexers and are sophisticated time allotment devices.

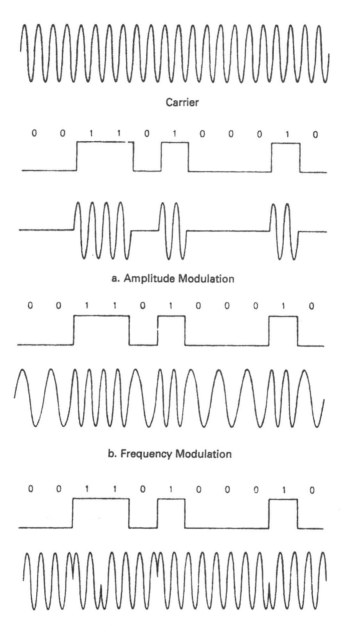

Carrier

a. Amplitude Modulation

b. Frequency Modulation

Figure 2-9 Three forms of modulation of the carrier frequency: (a) amplitude modulation, (b) frequency modulation, (c) phase modulation.

Figure 2-10 Time division multiplexing. The three signals are transmitted at different times.

2.6 SPECIAL APPLICATIONS OF TWISTED-PAIR WIRES

Special applications of twisted-pair wires require that various modifications be made to accommodate these applications. For example, for a high-electrical-noise environment, shielded twisted pair is constructed with each twisted pair being shielded by metal braid or metal mylar film. Applications of twisted-pair cables include workstation cables, distribution (feed) cables, riser (vertical distribution) cables, and cross-connected (jumper) cables.

Inside station cable has 2 to 12 pairs of wires for 1A2 mechanical key systems. The insulation is polyvinyl chloride (PVC) or Plenum rated (Teflon) for installation within the high-temperature areas of plenums and heat ducts.

Distribution cable is usually manufactured in 25 pairs and is common in 25 to 200 pairs. Special orders are available anywhere from 25 to 600 pairs. The insulation is usually gray PVC. However, Teflon-coated pairs can be ordered for resistance to high temperature. Table 2-8 gives a summary of two-wire cables: voice and data cables and twinaxial, all of which can be ordered in 200 to 3000 pairs.

Direct-bury cable (jelly-filled shielded) comes in 2 to 3600 pairs. The outer insulation on direct bury cable is impervious to water and chemicals in the soil. However, vermin, such as ground squirrels and gophers, will eat through the outer covering. This allows water and chemicals to attack the inner insulation and wire. Conduit should be used for any buried cabling.

Armored cable has a thick, strong outer jacket to protect the cable pairs.

The insulation on all cables is available in various colors for identification. All cables must meet the fire standards of the National Electrical Code (NEC), as discussed in Chapter 5.

Table 2-8 A matrix of the different types of twisted-pair cabling and applications

Description	MONCO CBL	UL Style	Conductor Material Stranding Diam (Inch)	Dielectric Material Diameter (Inch)	Shield Material % Coverage	Jacket Material Nom Diam (Inch)	Nom Vel. of Prop.	Nom Imped (ohms)	Nom Capac (pf/ft)	Nominal Attenuation MHz / dB/100'
Appletalk[1] Nonplenum	6242	2726	22 AWG 7/30 T.C.	PE .056	T.C. Braid 85% Coverage	PVC Snow Beige .183	66%	78	20.7	NA NA
Appletalk[1] Plenum	6228	NEC 725-38 (b) (3)	22 AWG 7/30 T.C.	FEP[3] .057	T.C. Braid 85% Coverage	FEP[3] Beige .184	70%	78	18.9	NA NA
78 Ohm Nonplenum CIM	6453	2092	20 AWG 7/28 T.C.	PE	T.C. Braid 93% Coverage	PVC Blue .242	65%	78	20.1	1 .6 / 10 2.1 / 20 3.0 / 50 5.0 / 100 7.5 / 200 11.0 / 400 16.0
78 Ohm Plenum	6292	NEC 725-38 (b) (3)	20 AWG 7/28 T.C.	FEP[3]	T.C. Braid 93% Coverage	Flouropolymer Blue Tint .220	70%	78	19.5	
78 Ohm RG 108-Type Factory Data Bus	6502	2582	20 AWG 7/28 T.C.	PE	Alum. Tape w/Drain + T.C. Braid 57% Coverage	PVC Blue .243	65%	78	20.1	1 .6 / 10 2.1 / 20 3.0 / 50 5.0 / 100 7.5 / 200 11.0 / 400 16.0
RG 22 B/U-Type	6503	2092	18 AWG 7/.0152 1 B.C. + 1 T.C.	PE .285	2 T.C. Braids 96% Coverage	PVC Black .420	65%	95	16.5	1 .3 / 10 .9 / 20 1.3 / 50 2.1 / 100 3.0 / 200 4.5 / 400 6.3

1 = Apple Computer trademark   2 = IBM Corp. trademark   3 = FEP—registered trademark of E.I. DuPont Co. Note: Consult Graybar for twinaxial variations not shown above

Table 2-8 A matrix of the different types of twisted-pair cabling and applications (Continued)

Description	MONCO CBL	UL Style	Conductor Material Stranding Diam (Inch)	Dielectric Material Diameter (Inch)	Shield Material % Coverage	Jacket Material Nom Diam (Inch)	Nom Vel. of Prop.	Nom Imped (ohms)	Nom Capac (pf/ft)	Nominal Attenuation MHz / dB/100'
100 Ohm IBM System 36[2] Nonplenum	6504	2498	20 AWG 7/28 1 B.C. + 1 T.C.	PE .240	T.C. Braid 95% Coverage	PVC Black .330	65%	100	15.6	1 → .4 10 → 1.1 20 → 1.5 50 → 2.5 100 → 4.1 200 → 6.4 400 → 10.2
100 Ohm IBM System 36[2] Plenum	6505	NEC 725-38 (b) (3)	20 AWG 7/28 1 B.C. + 1. T.C	FEP[3]	T.C. Braid 95% Coverage	Fluoropolymer Black Tint .282	66%	100	15.4	
124 Ohm Programmable Controller Type	5864	2092	25 AWG 7/33 T.C.	PE	Alum. Tape + T.C. Drain	PVC Blue .240	65%	124	.12.6	1 → .6 10 → 1.7 20 → 2.3 50 → 3.6 100 → 5.0 200 → 6.9 400 → 9.6
150 Ohm Factory Communication	6508	2668	22 AWG 19/34 T.C.	Foam Polypro	Alum. Tape + T.C. Drain	PVC Black .350	75%	150	9.0	1 → .3 10 → 1.3 20 → 2.3 50 → 3.0 100 → 4.3 200 → 6.2 400 → 8.8
124 Ohm Factory Communication	6506	2448	16 AWG Solid Bare Copper	Foam PE	Alum. Tape + T.C. Braid 95% Coverage	PVC Black .440	75%	124	10.9	1 → .27 10 → .89 20 → 1.3 50 → 2.0 100 → 2.9 200 → 4.0

1 = Apple Computer trademark   2 = IBM Corp. trademark   3 = FEP—registered trademark of E.I. DuPont Co. Note: Consult Graybar for twinaxial variations not shown above

Table 2-8 A matrix of the different types of twisted-pair cabling and applications (Continued)

Description	MONCO CBL	UL Style	Conductor Material Stranding Diam (Inch)	Dielectric Material Diameter (Inch)	Shield Material % Coverage	Jacket Material Nom Diam (Inch)	Nom Imped (ohms)	Nominal Attenuation MHz	dB/100'
TYPE 1 2 Data Pairs Nonplenum	6100-748C	Subject 13	22 AWG Solid Bare Copper	Foam PE .090	Alum. Foil Tape each Pair, plus overall, T.C. Braid 65% Coverage	PVC Black .380	150/4MHz	4 16	.66 1.37
TYPE 1 2 Data Pairs Plenum	6100-749C	NEC 725-38 (b) (3)	22 AWG Solid Bare Copper	Foam FEP .100	Alum. Foil Tape each Pair, plus Overall T.C. Braid 65% Coverage	Fluoropolymer Black Tint .365	150/4MHz	4 16	.66 1.37
TYPE II 2 Data Pairs 4 Voice Pairs Nonplenum	6100-739C	Subject 13	22 AWG Solid Bare Copper	2-Pair Foam FRPE 4-Pair Solid FRPE	2-Pair w/ individual Alum. Tape, plus overall T.C. Braid 65% Coverage	PVC Black .495	Data Pairs: 150/4MHz Voice Pairs: 100/4MHz	Data Pairs: 4 16 Voice Pairs: 1kHz 772/kHz	.66 1.37 .04 .51
TYPE II 2 Data Pairs 4 Voice Pairs Plenum	6100-738C	NEC 725-38 (b) (3)	22 AWG Solid Bare Copper	2-Pair Foam FRPE .100 4-Pair Solid FEP .045	2-Pair w/ individual Alum. Tape, plus overall T.C. Braid 65% Coverage	Fluoropolymer Black Tint .420	Data Pairs: 150/4MHz Voice Pairs: 100/4MHz	Data Pairs: 4 16 Voice Pairs: 1kHz 772/kHz	.66 1.37 .04 .51
TYPE 6 2-Pair Office Grade Data Cable	6100-743C	Subject 13	26 AWG 7/34 Bare Copper	Foam PE .070	Overall Alum. Tape plus T.C. Braid 65% Coverage	PVC Striated Black .327	150/4MHz	4 16	1.0 2.01

1 = Apple Computer trademark 2 = IBM Corp. trademark 3 = FEP—registered trademark of E.I. DuPont Co. Note: Consult Graybar for twinaxial variations not shown above

Table 2-8 A matrix of the different types of twisted-pair cabling and applications (Continued)

	Description	MONCO CBL	UL Style	Number of Pairs	Conductor Material Stranding Diam (Inch)	Dielectric Material Diameter (Inch)	Colors	Jacket Material Nom Diam (Inch)	Shield Material % Coverage	Nom Imped (ohms)
	Usernet Pair Nonplenum	6734	2509	1	20 AWG 10/30 T.C.	PVC .072	Black + Red	PVC Gray .190	NA	NA
	Usernet Pair Nonplenum Shielded	5624	2092	1	20 AWG 7/28 T.C.	PE .068	Black + Clear	PVC Gray .204	Alum. Tape plus 20 AWG Drain	NA
	Usernet Pair Plenum	6507	NEC 725-38 (b) (3)	1	20 AWG 7/28 T.C.	Tefzel^2 .066	Black + Red	Tefzel^2 Clear .172	NA	NA
	Omninet^3 Pair Nonplenum	6408	2919	1	24 AWG 7/32 T.C.	Foam Polypropylene .064	Black + Red	PVC Gray .201	NA	135
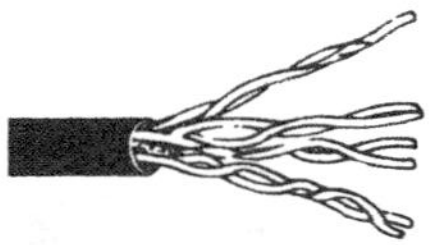	IEEE802.3 1-Base-5	6513	20245	4	24 AWG 7/32 T.C.	Semi-rigid PVC .039	Telephone Standard	PVC White .205	NA	95
	IEEE802.3 1-Base-5 Plenum	0586-03	NEC 725-38 (b) (3)	4	24 AWG 7/32 T.C.	FEP .0075	Telephone Standard	Kynar	NA	95
	Type 3 Token Ring	6512	NA	4	24 AWG Solid Bare	Semi-rigid PVC .033	Telephone Standard	PVC Gray .170	NA	100

1 = Trademark of Sperry Corp.—Usernet   2 = Trademark of E.I. DuPont Co.—Tefzel   3 = Trademark of Corvus Systems—Omninet
Note: Please consult Graybar for designs not shown here.

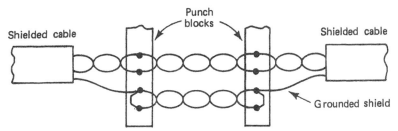

Figure 2-11   An example of a ground connection for a twisted pair between punch-down blocks.

2.7 TWISTED-PAIR CABLE TERMINATION

Regardless of the type of twisted-pair cable that is used, each wire must be terminated at the source, at the destination, and at splices in long cables. Cables can be purchased with a variety of connectors for almost any application. These same connectors can be purchased and installed on cables. The termination of a sampling of equipment is as follows:

- Station jacks: 2 to 12 pairs, with 3 to 6 pairs the most common
- Duplex jacks (two jacks): usually have a voice and data combination
- Surface mount connectors
- Flush mount connectors
- RJ11[†] with six positions and four contacts
- RJ45 with six positions and six contacts
- RJ45 with eight positions and eight contacts

Multifunction custom-built connectors and terminals are available in the push-on and screw type. Some equipment manufacturers, such as IBM, AT&T, and Digital Equipment, require custom connectors. A great variety of termination connectors is available from distributors, some of which are listed in the Vendor section of this book.

Grounding of the cable shield is an important part of the termination of shielded twisted-pair cable. Special grounding techniques will be covered later in this chapter. However, there are techniques that apply specifically to twisted pair—for example, the continuation of a ground through a pair of punch-down blocks (Figure 2-11). The ground from each cable is connected to a twisted pair of wires that completes the connection. The twist in the connection prevents any electromagnetic pick-up in the ground extension between the blocks.

When cable enters or exits a building, a lightning arrestor must be installed between the lines and ground. The device acts as an open circuit unless a very large voltage, such as a lightning bolt, surges between the wires and ground. A large voltage surge causes a short circuit to

---

† RJ jacks are jacks that are registered with the FCC. There are more than 30 types of RJ jacks used in telephone and telecommunication technology.

Figure 2-12 An example of a lightning arrestor
that would be used on a twisted-pair line. (*Source:*
Courtesy Anixter Brothers, Inc.)

ground, which gives a low-resistance path across which the energy can dissipate. Figure 2-12 depicts one such type of lighting protector.

2.8 DISTRIBUTION FRAMES

When a system installation requires many cables, the cables are usually run from a *main distribution frame* (MDF). The MDF is a point at which the cables can be added to, deleted from, or rerouted in the system. The MDF (Figure 2-13) is usually located in a room or area called a closet. A large system may have auxiliary or *terminated distribution frames* (TDFs).

A typical AT&T MDF has 66 blocks vertically mounted on 89 B brackets. These frames can be purchased individually or per standard manufactured sizes, with 25 pairs and 50 pair-split blocks being the most common. An example of a telephone termination block is shown in Figure 2-14.

Special application blocks are available. For example, AT&T 110 PDS equipment is horizontally orientated and high density, with a complete line of blocks, mounting equipment, and labeling. Auxiliary equipment, such as patch cords, racks, and patching tools, is also available in this line.

Another line of MFD or TDF equipment is BIX[‡] blocks. This line is horizontally oriented and high density. A complete line of products, such as blocks, mounting brackets, labels, and patch cords, and so on, is available. Special termination tools are necessary for installation of the cables.

2.9 EXISTING CABLE SYSTEMS AND COMPATIBILITY

Most businesses have an existing telephone cabling system in place, which can be transformed into a LAN with some data processing equipment. The communication manager must ask the following questions before deciding to use in-place telex wiring:

• Can the existing system be incorporated into a new plan?
• Is the old equipment compatible with the new system?

---

‡ BIX ia a registered trade name for Northern Telecom System of an in-building termination and cross-connected system.

Figure 2-13 A photograph of a main distribution frame. (*Source:* Courtesy Nevada Western Corp.)

- Are current staff members capable of installing new lines and new equipment?
- Should the LAN use a telephone system, a PBX, a key system, or DSL?
- Should the old twisted pair be replaced or connected to coax or fiber, and what type of connection devices are necessary?
- Are the current test instruments sufficient for testing and certifying the new system?

With the magnitude of equipment providers available, and their effort for compatibility and seamless connections, the answer to all of the preceding questions is probably "yes."

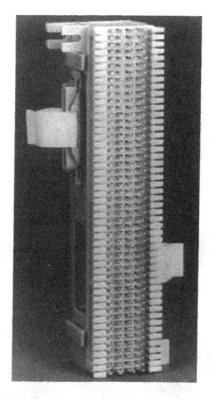

Figure 2-14 An example of a telephone termination block. (*Source:* Courtesy of Nevada Western Corp.)

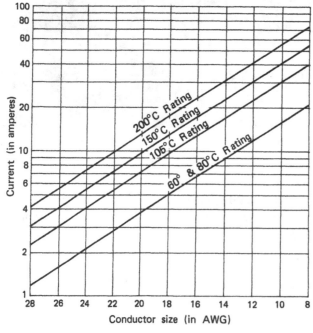

No. of conductors*	Factors
1	1.6
2-3	1.0
4-5	.8
6-15	.7
16-30	.5

* Do not count shields unless used as a conductor

Figure 2-15 Current versus conductor size for different temperatures. (*Source:* Belden Inc., Electronics Division)

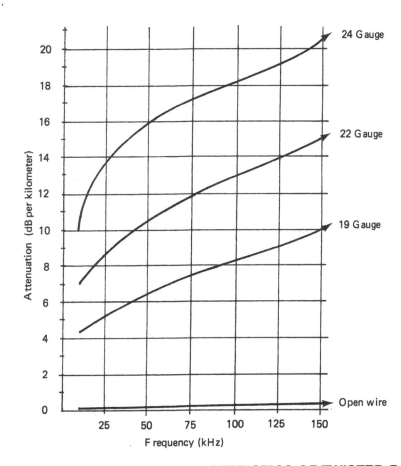

Figure 2-16 Graph of frequency versus attenuation in decibels for different sizes of conductors. (*Source:* Belden Inc., Electronics Division)

2.10 ELECTRICAL CHARACTERISTICS OF TWISTED-PAIR CABLING

In this section we will discuss the electrical characteristics particular to twisted-pair cables. Figure 2-15 is a chart of the continuous current ratings of conductors at four different temperatures for different sizes of wire. For example, a #24-gauge wire has a rating of approximately 1.8 amperes at temperatures between 60 and 100°C.

Figure 2-16 depicts the attenuation in decibels per 100 feet for three different types of twisted-pair packaging. The important consideration here is that the decibel loss increases significantly as the frequency of the signal on the cable increases. Note that a 10-dB power loss reduces signal power to one-tenth its original value and voltage to approximately one-third its original value.

Figure 2-17 gives the characteristics of typical twisted-pair wiring.

Figure 2-18 gives the electrical characteristics of twinaxial cable. Figure 2-18a is an attenuation versus frequency chart for the cable. Figure 2-18b is a chart of the rise time of transmitted pulses versus the transmission distance. You may remember from Chapter 1 that the capacitance

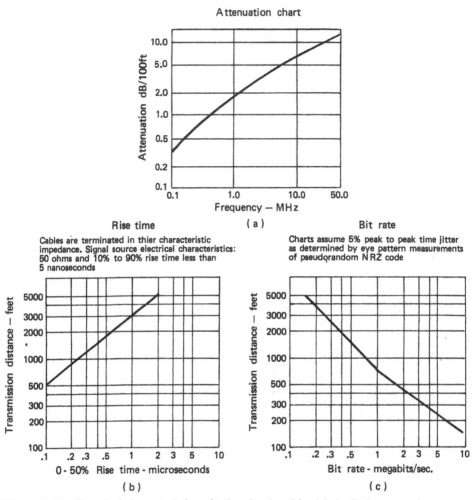

Figure 2-17 Electrical characteristics of twisted-pair cable: (a) decibels versus frequency, (b) pulse rise time versus transmission distance, and (c) bit rate versus transmission distance. (*Source:* Belden Inc., Electronics Division)

of a cable is directly proportional to the length of the cable, and the greater the capacitance the greater the rise time of the digital pulses transmitted along the cable. Figure 2-18c is a chart of the allowable bit rate of data pulses versus the transmitted distance of the signal. From the chart we see that a transmission distance of 1000 feet would allow a bit rate of 500,000 bits per second. The maximum bit rate of a cable is dependent to a great degree on the resistance of the wires and the capacitance of the cable. Note that the ratings given in the charts are for transmission distance and not the length of the cable, which would be twice as much as the transmission distance. When calculating the transmission distance, the length of all the jumper wires connecting data processing equipment to the cable must be added to the distance. Another consideration

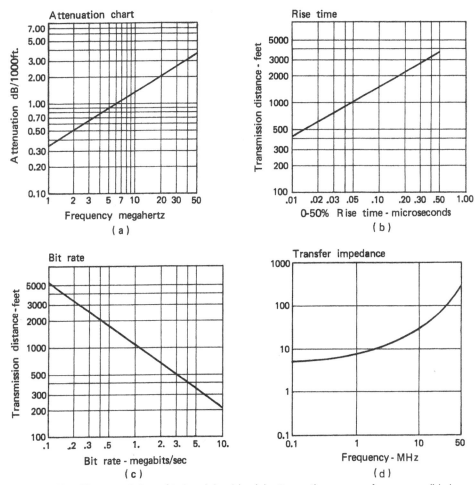

Figure 2-18 Characteristics of twinaxial cable: (a) attenuation versus frequency, (b) rise time versus transmission distance, (c) bit rate versus transmission distance, (d) frequency versus transfer impedance.

when making connections to twisted-pair cabling is the characteristic impedance of the cable. Figure 2-18d is a plot of the frequency versus transfer impedance for twinaxial cable.

When a coaxial cable of a certain impedance is connected to twisted pair, data processing equipment, or other coaxial cable of a different impedance, a matching device called a balun must be used to prevent signal loss. A balun is a transformer device that matches the impedance of one cable to cables or devices that have a different characteristic impedance. The balun electrically isolates the lines and makes each appear to be matched to the correct impedance, and thereby, prevents serious signal loss.

Shielding is effective in preventing cross talk between twisted pairs. To be most effective the shielding must extend to the connectors. Figure 2-19 depicts the effects of shielding on the

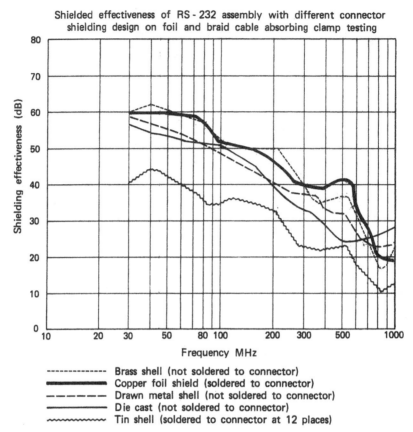

Figure 2-19   Shielding effectiveness versus frequency for a twisted pair.
(*Source:* Belden Inc., Electronics Division)

shielding ability of an RS-232-type connector. Notice that the shielding is generally less effective for all types of materials as the frequency of the outside interference. For example, in the case of copper foil shielding, there is an increase of almost 10 dB as the frequency increases from 50 MHz to 60 MHz. This means that the shielding is only one-eight as effective at the higher frequency.

2.11 FLAT CABLE AND RIBBON CABLE

Flat cabling lends itself to many applications due to its thin and rugged construction. For example, the thin construction allows it to be installed under carpets without creating a ridge (which might be a safety hazard or cause excessive wear of the carpet). An example of flat cable is shown in Figure 2-20.

Ribbon cabling is designed to be placed in an environment that will produce less wear than that for flat cabling. For example, flat cabling is excellent for interconnection of modules within

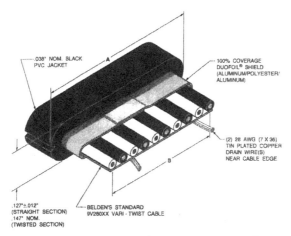

.038" NOM. BLACK
PVC JACKET

100% COVERAGE
DUOFOIL® SHIELD
(ALUMINUM/POLYESTER/
ALUMINUM)

(2) 28 AWG (7 X 36)
TIN PLATED COPPER
DRAIN WIRE(S)
NEAR CABLE EDGE

.127"±.012"
(STRAIGHT SECTION)
.147" NOM.
(TWISTED SECTION)

BELDEN'S STANDARD
9V280XX VARI - TWIST CABLE

Figure 2-20   Flat cable. (*Source:* Belden Inc., Electronics Division)

equipment such as a PC. The nature of the cabling prevents the wires from twisting and allows the flexibility necessary when installing and removing printed circuit boards. Flat cabling has an added advantage over individual wiring in that connectors are easier to install. There are some drawbacks to flat cabling. It usually is comprised of small wires and has greater cross talk and attenuation than twisted pair. It may require impedance-matching transformers to reduce cross talk. These factors should be considered before selecting any type of flat cable.

2.12   TOOLS FOR INSTALLATION OF TWISTED-PAIR WIRES

The installation of each type of cabling requires special tools. In this section we will discuss some of those tools required for the installation of twisted-pair cables.

The diagonal pliers depicted in Figure 2-21 are typical of electrical wire cutters. Diagonal pliers are available in many different sizes to handle various sizes of wires. The electrical pliers shown in Figure 2-22 can be used for cutting wires and for holding wires or parts in place. Electrical or electrician's pliers are available in several sizes to allow cutting of various sizes of wire.

The electrical wire stripper-cutter crimping tool shown in Figure 2-23 can serve several functions:

- The tips can be used as wire cutters.
- The inside of the handles can be used as wire strippers.
- The holes in the handles are threaded so that screws can be cut to length without ruining the threads.
- The blades are designed for connector crimping.

Figure 2-24 shows a precision wire stripper that, when used properly will prevent the nicking of a wire. The wire is inserted into the proper slot (by AWG #10 to #20) and the handles are

Figure 2-21   Diagonal pliers used as wire cutter.

Figure 2-22   Electrical pliers.

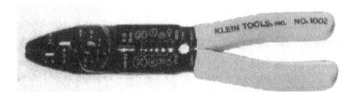

Figure 2-23   Wire stripper/crimper.

squeezed. When the handles are released slowly, the insulation is removed. The tips of this tool are replaceable for application with smaller wire sizes. This tool is preferable over the wire stripper shown in Figure 2-23, especially for small solid copper conductors.

The stripper shown in Figure 2-25 uses battery-heated tips to melt plastic insulation from copper wire. It can also be used to strip the coating from glass fiber-optic cable. Figure 2-26 depicts another small precision stripper and cutter that can be carried easily in a pocket or tool pouch. The drawing illustrates the proper use of the tool.

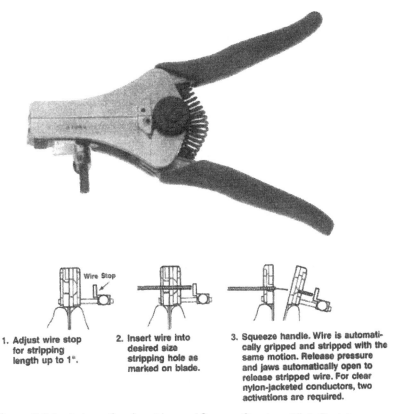

1. Adjust wire stop for stripping length up to 1".

2. Insert wire into desired size stripping hole as marked on blade.

3. Squeeze handle. Wire is automatically gripped and stripped with the same motion. Release pressure and jaws automatically open to release stripped wire. For clear nylon-jacketed conductors, two activations are required.

Figure 2-24   Automatic wire stripper. (*Source:* Courtesy KleinTools)

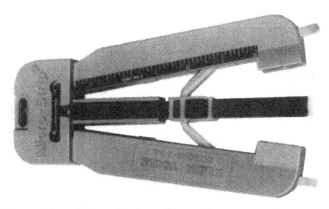

Figure 2-25   Thermal stripper for small copper conductor or fiber-optic cable. (*Source:* Courtesy KleinTools)

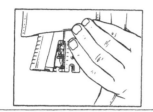

Automatic adjustment to cable area

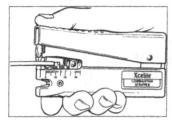

Built-in cutter

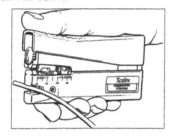

Figure 2-26 Wire cutter/stripper designed for the wiring technician.

Still another small stripper for insulation removal and wire stripping is shown in Figure 2-27. Each conductor of the multiconductor cable is prepared and inserted into the punch-down terminals of the connector. The wires of the cable in Figure 2-27 are "punched down" or pressed

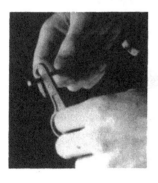
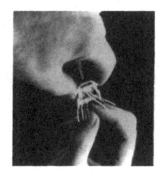

(a)   (b)   (c)

Figure 2-27 Installing multiple strand conductors: (a) stripping the cable, (b) inserting the conductors into a connector, (c) finished connector. (*Source:* Courtesy Nevada Western Corp.)

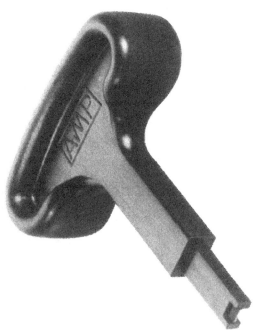

Figure 2-28   Punch-down tool. (*Source:* Courtesy Amp Inc.)

into the slots in the terminals by a punch-down tool such as shown in Figure 2-28. The punch-down tool is most often used to force single-strand #22 or #24 copper wire into a telephone block. Figure 2-29 depicts a type of connector that requires no punch down. The wires are inserted into the pins as shown in Figure 2-29a and clipped to length with diagonal pliers. The cover of the connector is then forced into place with channel lock pliers. The cover acts as a punch-down tool, forcing the edges of the connectors through the insulation and making contact.

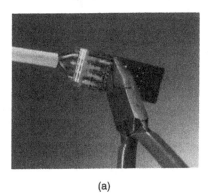
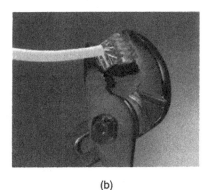

(a)   (b)

Figure 2-29   Modular connector: (a) inserting and cutting the wires, (b) locking the connector. (*Source:* Courtesy Progressive Electronics, Inc.)

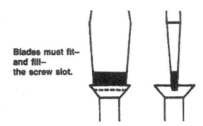

Blades must fit—
and fill—
the screw slot.

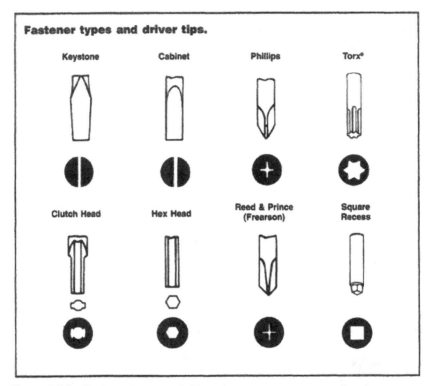

Figure 2-30   Fastener types and driver tips.

One of the basic tools for installation of wire and repair of equipment and systems is the screw fastener or screwdriver. Everyone knows how to use this tool, but not everyone uses it correctly. There are many sizes of screw heads, and many different screw slots in the heads. Figure 2-30 depicts the common screw slot types. Each slot requires a different tool. Furthermore, each slot size requires the correct size of tip. Therefore, the professional requires a number of screwdrivers in a variety of sizes, lengths, and types. The wrong tool—incorrect slot type, slot size, or driver length—can cause wasted time and possible trouble.

There are dozens of specialized tools for wire handling and installation. The professional must keep abreast of the latest time-saving innovations. However, the technician must be prepared with a basic set of tools that will meet most situations. A battery-operated drill/screw-driver should be added to any basic set of tools as a time- and energy-saving tool.

2.13 ADVANTAGES OF TWISTED-PAIR WIRING

Some of the advantages of twisted-pair wiring over other types of cabling are as follows:

1. Twisted pair is in place as telephone cabling in all but the newest installations.
2. Twisted pair can be utilized with coax and fiber cabling.
3. Distribution cable can be used and home runs usually will not be necessary.
4. New technology, such as DSL (Chapter 7), has greatly extended the data speed and bandwidth of twisted pair and thereby the applications.
5. Although fiber will be the choice of the future, copper will rule for many more years.

2.14 SUMMARY

Twisted-pair wiring is, and will be for years, the most often used telecommunication medium. A driving technology is to use new techniques of transmission to maximize the transmission rate of twisted pair. New inventions and new transmission techniques have increased the bandwidth capability of twisted pair far beyond what was thought the maximum a few years past.

QUESTIONS

1. What are the limitations of twisted-pair cables?
2. What are the advantages of twisted-pair cables?
3. What is the difference between #24 and #22 twisted-pair wire?
4. What is the purpose of grounding the shield on a cable?
5. What is the advantage of flat cabling over ordinary cabling?
6. What is the purpose of a balun?

CHAPTER   3

Coaxial Cable

3.1 INTRODUCTION

A coaxial cable (commonly called coax) is comprised of a center conductor and an outer shielding conductor covered by an outer insulating jacket (Figure 3-1). The center conductor can be a
single or stranded wire. Single conductors have less resistance than stranded conductors but are
less flexible and can kink unless care is used in the installation. Some coaxial cables are constructed with a strain relief in the form of a small-diameter fiber or plastic line wrapped around
the center conductor. Coax cables that are used as patch panels or service bundles should have
this strain relief.

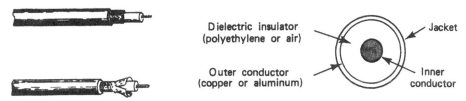

Figure 3-1   Construction details of coaxial cable: an older type of braided shield and a
TNC type with a solid shield.

The center conductor of a coax is surrounded by an insulating material, called a **dielectric**,
which is in turn surrounded by an outer metallic shield and jacket. The magnetic shield serves
both as the return path for the electrical energy and as a shield for the center conductor against
electromagnetic pick-up and cross talk. It also limits radiation of the signal from the center conductor. The tough outer insulating jacketing material that covers the metallic shield is usually
some form of rubber. When a very large number of signals is to be transported over the same

Figure 3-2 A coaxial cable with
two inner conductors, packaged to
form two coaxial cables.

route but over individual coaxial cables, coax can be bundled into a multicable jacket for ease of
installation.

3.2 CHARACTERISTICS AND CONSTRUCTION OF COAXIAL CABLE

Coaxial cable is also constructed with two center conductors, each a coaxial cable in itself. The
two coaxs are insulated from each other, surrounded by a dielectric, enclosed by a magnetic
shield, and surrounded by an outer insulator (Figure 3-2). This packaging, called *twinaxial
cable*, forms two coaxial cables in a small diameter.

The twinaxial transmission line cable offers low-loss signal transmission that is unaffected
by outside noise from electromagnetic sources such as cross talk, motor noise, and magnetic
emission from florescent lights. Figure 3-3 summarizes the characteristics of typical coaxial
cable. Figure 3-3a gives the attenuation in decibels versus frequency in megahertz. Figure 3-3b
gives the bit rate versus transmission distance, and Figure 3-3c gives the pulse rise time versus
transmission distance. The characteristics given in these figures may be compared to those given
for twisted pair in Chapter 2.

There are several types of coaxial cables, each with the same general construction but with
somewhat different electrical characteristics. Figure 3-1 denoted the two basic types of construction:

1. solid metallic shield
2. braided metallic shield

The following are the types of coaxial cable and their respective characteristic
impedances:

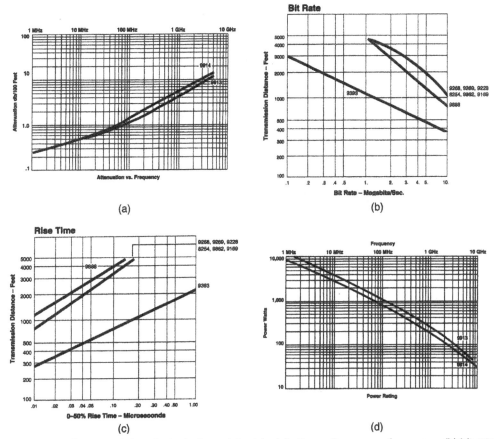

Figure 3-3 Characteristics of a typical coaxial cable: (a) attenuation versus frequency, (b) bit rate versus transmission frequency, (c) rise time versus distance, (d) power rating versus frequency. (*Source:* Belden Electronics Division)

RG 6*	75 Ω	18 AWG center conductor
RG 8	50 Ω	18 AWG center conductor
RG 11	75 Ω	14 AWG center conductor
RG 58	50 Ω	14 AWG center conductor
RG 59	75 Ω	22 AWG center conductor
RG 62	92 Ω	22 AWG center conductor

Ethernet trunk cable is usually of the 50-Ω impedance type, such as RG 8 or RG 58. Video cable is usually of the 75-Ω type such as RG 59. Coaxial cable is known for its wide bandwidth, less electromagnetic radiation, less cross talk, and greater security than twisted pair. Ethernet is a

* RG indicates coaxial cable. For example, RG 58 is a 50-ohm coaxial cable used for Thinnet®.

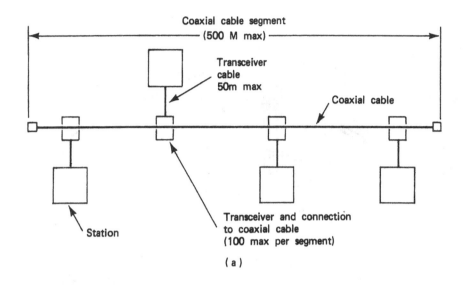

(a)

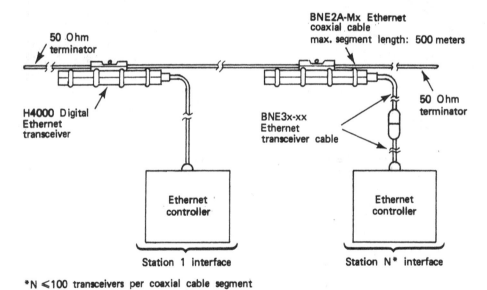

*N ≤100 transceivers per coaxial cable segment

(b)

Figure 3-4   Ethernet network system: (a) basic system, (b) transceiver drop connection.

local area network developed by Xerox and then sponsored by Digital Equipment Corp. (DEC) and Intel Corp. to become an industry standard. In an Ethernet system the receive and send signals are all transported along a single coaxial cable (Figure 3-4a) and coupled to various data communication equipment by an Ethernet transceiver called a *drop point* (Figure 3-4b). The

Figure 3-5 Barrel-type connector
used to connect coaxial cables.
(*Source:* Courtesy TYCO
Electronics Corporation)

maximum cable length is approximately 500 m. Ethernet coaxial cable is available in various lengths, which can be connected with barrel connectors (Figure 3-5). The barrel connector causes very little signal attenuation.

Although all coaxial cable has basically the same construction, there are a number of types manufactured to withstand specific environmental conditions. The size of the inner conductor, the insulation dielectric material, and the outer shield material used result in cables of different characteristic impedances. Table 3-1 summarizes the types of coaxial cable available.

We note from Table 3-1 that the characteristic impedance for coax ranges between 50 and 95 Ω. Cables are available that contain twisted pair, twinaxial, and/or coax. Table 3-2 summarizes one manufacturer's inventory.

3.3 COAXIAL CABLE CONNECTORS AND TERMINATIONS

Whenever possible, coaxial cable should be purchased with the connectors attached. However, this is seldom possible for long cable runs and points of cable drops where outputs are taken for data processing equipment. These connections will require that cable adapters be installed by the wiring technician.

Table 3-1   Types of coaxial cables

- **Cable RG 59/U-1 (105 482 624)** is a 75 ohm coaxial cable with a 22 AWG (7x30) center conductor, a foamed polyethylene dielectric, a bare copper braid (min. 95% coverage) outer conductor and a PVC jacket. (Similar to RG 59/U type.) UL style 1354.

- **Cable RG 59/U-1A (105 521 561)** is a 75 ohm coaxial cable with a 22 AWG (7x30) center conductor, a foamed polyethylene dielectric, a bare copper braid (min. 95% coverage) outer conductor and a PVC jacket. (Similar to RG 59/U type.) UL style 1354, UL Listed Type CL2.

- **Cable RG 59/U-2 (105 482 632)** is a 75 ohm coaxial cable with a 22 AWG copper covered steel center conductor, a polyethylene dielectric, a bare copper braid (min. 80% coverage) outer conductor and a PVC jacket. (Similar to RG 59/U type commercial.) UL style 1354.

- **Cable RG 59/U-2A (105 521 579)** is a 75 ohm coaxial cable with a 22 AWG copper covered steel center conductor, a polyethylene dielectric, a bare copper braid (min. 80% coverage) outer conductor and a PVC jacket. (Similar to RG 59/U type commercial.) UL style 1354. UL Listed Type CL2 per 1987 NEC.

- **Cable RG 59/U-5 (105 482 665)** is a 75 ohm plenum coaxial cable with a 22 AWG copper covered steel center conductor, an FEP dielectric, a bare copper braid (min. 95% coverage) outer conductor and an FEP jacket. (Similar to RG-59/U type.) UL Listed Type CL2P per 1987 NEC.

- **Cable RG 62 A/U-1 (105 482 723)** is a 93 ohm coaxial cable with a 22 AWG copper covered steel center conductor, an air dielectric polyethylene dielectric, a bare copper braid (min. 95% coverage) outer conductor and a PVC jacket. (Similar to RG 62 A/U type.)

- **Cable RG 62 A/U-1A (105 521 660)** is a 93 ohm coaxial cable with a 22 AWG copper covered steel center conductor, an air dielectric polyethylene dielectric, a bare copper braid (min. 95% coverage) outer conductor and a PVC jacket. (Similar to RG 62 A/U type.) UL Listed Type CL2 per 1987 NEC.

- **Cable Ethernet™-1 (105 482 798)** is a 50 ohm coaxial cable with a 0.0855 AWG solid tinned copper center conductor, a foamed polyethylene dielectric, a foil shield bonded to dielectric, a tinned copper braid (min. 92% coverage) a foil shield, a tinned copper braid (min. 92% coverage) and a PVC jacket. (Similar to Ethernet Type.) UL style 1478 DEC approved. "DEC NET." Xerox specifications/IEEE 803.

- **Cable Ethernet-1A (105 538 037)** is a 50 ohm coaxial cable with a 0.0855 AWG solid tinned copper center conductor, a foamed polyethylene dielectric, a foil shield bonded to dielectric, a tinned copper braid (min. 93% coverage) a foil shield, a tinned copper braid (min. 90% coverage) and a yellow PVC jacket. (Similar to Ethernet Type.) UL style 1478. UL Listed Type CL2 per 1987 NEC. DEC approved. "DEC NET." Xerox specifications/IEEE 803.

- **Cable Ethernet-2 (105 482 806)** is a 50 ohm plenum coaxial cable with a 0.0855 AWG solid tinned copper center conductor, a foamed FEP dielectric, a foil shield, a tinned copper braid (min. 93% coverage), a foil shield, a tinned copper braid (min. 90% coverage) and an FEP jacket. (Similar to Ethernet.) UL Listed Type CL2P per 1987 NEC. DEC approved. Xerox specifications/IEEE 803.

The two types of connectors utilized are BNC (Bayonet Neill-Concelman) and TNC (Twist Neill-Concelman). The BNC is a bayonet type developed for easy installation of video-type cable (cable with a solid shield). Examples of these two types of connectors are shown in Figure 3-6. The BNC and TNC notations indicate the type of locking of the collar. The connector shown in Figure 3-6c is an economical plastic type. The plastic collar protects personnel from possible electrical shock. The collars are available in six colors to help identify individual cables when both ends are not within view.

Table 3-2 Composite building cables

Description	MONCO CBL	UL Style	Conductor Material Stranding Diam (Inch)	Dielectric Material Diameter (Inch)	Shield Material % Coverage	Jacket Material Nom Diam (Inch)	Nom Vel. of Prop.	Nom Imped (ohms)	Nom Capac (pf/ft)	Nominal Attenuation MHz dB/100'
RG 59/U type plenum	6460	NEC 725-38 (b) (3)	.032 Copper Covered Steel	Foam FEP[1] .146	Bare Copper Braid 95% Coverage	Fluoropolymer Black .205	81%	75	16.0	50 — 1.8; 100 — 2.6; 200 — 3.8; 500 — 6.2
RG 59/U type nonplenum	6489	1354	.025 Bare Copper Covered Steel	PE .146	Bare Copper Braid 95% Coverage	PVC Black .242	65.5%	73	21.3	50 — 2.4; 100 — 3.4; 200 — 4.9; 400 — 7.1; 700 — 9.5; 900 — 10.9; 1000 — 12.0
RG 59/U type plenum	6490	NEC 725-38 (b) (3)	.025 Bare Copper Covered Steel	FEP[1] .135	Bare Copper Braid 95% Coverage	Fluoropolymer Black Tint .206	69%	75	19.6	100 — 3.4; 200 — 4.9; 400 — 7.1
RG 59/U type nonplenum cellular	6234	1354	22 AWG 7/30 Bare Copper	Foam PE .146	Bare Copper Braid 95% Coverage	PVC Black .242	75%	75	18	50 — 2.1; 100 — 3.0; 200 — 4.5; 400 — 6.6; 700 — 8.9; 900 — 10.1; 1000 — 10.9
RG 59/U type plenum cellular	6492	NEC 725-38 (b) (3)	22AWG 7/30 Bare Copper	Foam FEP[1] .146	Bare Copper Braid 95% Coverage	Fluoropolymer Black Tint .218	81%	75	16.7	50 — 2.1; 100 — 3.0; 200 — 4.5; 400 — 6.6; 900 — 10.1
Dual RG 59/U type nonplenum	6491	20063	.023 Bare Copper Covered Steel	PE .146	Bare Copper Braid 95% Coverage	PVC Black .238 × .478	65.5%	75	20.7	100 — 3.4; 200 — 5.1; 400 — 7.5; 700 — 11.4; 900 — 12.0; 1000 — 12.7
Dual RG 59/U type plenum	6454	NEC 725-38 (b) (3)	.023 Bare Copper Covered Steel	FEP[1] .134	Bare Copper Braid 95% Coverage	Fluoropolymer Clear .236 × .442	69%	75	19.6	100 — 3.5; 200 — 5.1; 400 — 7.5; 700 — 11.4; 900 — 12.0; 1000 — 12.7

Table 3-2 Composite building cables (Continued)

Description	MONCO CBL	UL Style	Conductor Material Stranding Diam (Inch)	Dielectric Material Diameter (Inch)	Shield Material % Coverage	Jacket Material Nom Diam (Inch)	Nom Vel. of Prop.	Nom Imped (ohms)	Nom Capac (pf/ft)	Nominal Attenuation MHz	B/100'
triax			Covered Steel		Coverage w/ PE insulation between braids	.315				200 300 400 900	3.8 4.8 5.6 8.4
RG 59/U type plenum triax	6494	NEC 725-38 (b) (3)	.032 Bare Copper Covered Steel	Foam FEP[1] .140	2 Bare Copper Braid 96% Coverage w/ Fluoropolymer insulation between braids	Fluoropolymer Black Tint .262	81%	75	16.7	50 100 200 500 900	1.8 2.5 3.6 6.0 8.6
RG 62 A/U type nonplenum	5770	1478	22 AWG Solid Bare Copper Covered Steel	Semi-solid PE .146	Bare Copper Braid 90% Coverage	PVC Black .242	80%	93	13.7	400	8.0
RG 62 A/U type plenum cellular	5727-1	NEC 725-38 (b) (3)	22 AWG Solid Bare Copper Covered Steel	Foam FEP[1] .146	Bare Copper Braid 90% Coverage	Fluoropolymer White Tint .225	80%	93	14.5	400	8.0
RG 8/U type nonplenum	6495	1354	11 AWG 7/19 Bare Copper	Foam PE .285	Bare Copper Braid 97% Coverage	PVC Black .405	78%	50	26.1	50 100 200 400 700 900 4000	1.2 1.8 2.7 4.2 5.8 6.7 18.0
RG 8/U type plenum	6496	NEC 725-38 (b) (3)	11 AWG 7/19 Bare Copper	Foam FEP[1] .285	Bare Copper Braid 97% Coverage	Fluoropolymer Black Tint .365	81%	50	25.1	50 100 200 400 700 900 4000	1.2 1.8 2.7 4.2 5.8 6.7 18.0
RG 11/U type nonplenum	6497	1354	.064 Bare Copper	Foam PE .285	Bare Copper Braid 95% Coverage	PE .405	78%	75	17.4	50 100 200 500 900	1.0 1.5 2.2 3.7 5.2

[1]FEP = Registered trademark of E. I. Du Pont de Nemours & Co.

Note: Please consult Graybar for coaxial variations not shown above.

Table 3-2 Composite building cables (Continued)

Description	MONCO CBL	UL Style	Conductor Material Stranding Diam (Inch)	Dielectric Material Diameter (Inch)	Shield Material % Coverage	Jacket Material Nom Diam (Inch)	Nom Vel. of Prop.	Nom Imped (ohms)	Nom Capac (pf/ft)	Nominal Attenuation MHz	dB/100'
RG 11/U type nonplenum	6498	1354	.064 Bare Copper	Foam PE .285	Alum/Tape + T.C. Braid 61% Coverage	PVC .405	78%	75	17.4	50 100 200 500 900	
RG 11/U type plenum	6499	NEC 725-38 (b) (3)	.064 Bare Copper	Foam FEP[1] .285	Alum/Tape + T.C. Braid 61% Coverage	Fluoropolymer .363	80%	75	16.9	50 100 200 500 900	
RG 11/U type nonplenum triax	6500	1354	.064 Bare Copper	Foam PE .285	2 Bare Copper Braid 96% Coverage w/ PE insulation between braids	PE .475	78%	75	17.4	50 100 200 300 400 900	
RG 11/U type plenum quad	6501	NEC 725-38 (b) (3)	.064 Bare Copper Covered Steel	Foam FEP[1] .280	2 Alum/Tapes + 2 T.C. Braids	Fluoropolymer .389	82%	75	16.5	400	
Ethernet[2] nonplenum	5688	1478 60°C	.0855 Bare Copper	Foam PE .247	Alum/Tape + T.C. Braid 90% Min. Cov. + Alum/Tape + T.C. Braid 90% Min. Cov.	PVC Yellow .405	78%	50	26.0	5 10	

[1]FEP = Registered trademark of E. I. Du Pont de Nemours & Co.

Table 3-2 Composite building cables (Continued)

Description	MONCO CBL	UL Style	Conductor Material Stranding Diam (Inch)	Dielectric Material Diameter (Inch)	Shield Material % Coverage	Jacket Material Nom Diam (Inch)	Nom Vel. of Prop.	Nom Imped (ohms)	Nom Capac (pf/ft)	Nominal Attenuation MHz	dB/100'
Ethernet[4] plenum	5713-1	NEC 725-38 (b) (3) 125°C	.0855 Bare Copper	Foam FEP[1] .245	Alum/Tape + T.C. Braid 90% Min. Cov. + Alum/Tape + T.C. Braid 90% Min. Cov.	Kynar Flex[3] Orange .375	78%	50	26.0	5 / 10	.38 / .53
Thin-net nonplenum IEEE 802.3 10-Base-2	6417	1354 80°C	19/.0072 Tin Copper	Solid PE .116	T.C. Braid—95% Coverage	PVC Black .195	66%	50	30.8	5 / 10	.99 / 1.4
Thin-net nonplenum IEEE 802.3 10-Base-2	6418	1354 80°C	19/.0072 Tin Copper	Foam PE .094	Alum/Tape + T.C. Braid 95% Coverage	PVC Black .185	78%	50	26.0	5 / 10	.99 / 1.4
Thin-net plenum IEEE 802.3 10-Base-2	6420	NEC 725-38 (b) (3) 125°C	19/.0072 Tin Copper	Solid FEP[1] .110	Alum/Tape + T.C. Braid 95% Coverage	Kynar Flex[3] Black Tint .174	66%	50	30.8	5 / 10	.99 / 1.4
Thin-net plenum IEEE 802.3 10-Base-2	6419	NEC 725-38 (b) (3) 125°C	19/.0072 Tin Copper	Foam FEP[1] .090	Alum/Tape + T.C. Braid 95% Coverage	Kynar Flex[3] Black Tint .160	78%	50	26.0	5 / 10	.99 / 1.4

[1]FEP = Registered trademark of de Nemours & Co.

[2]Trademark of Xerox Corp.

[3]Kynar—Registered trademark of Penwalt Corporation

Note: All constructions shown can be altered to satisfy individual customer requirements. For instance, plenum jacketing materials can be FEP, Halar, etc. Consult Graybar with specific requirements.

Table 3-2 Composite building cables (Continued)

Description	MONCO CBL	UL Style	Conductor Material Stranding Diam (Inch)	Dielectric Material Diameter (Inch)	Shield Material % Coverage	Jacket Material Nom Diam (Inch)	Nom Imped (ohms)	Nominal Attenuation MHz / dB/100'
Figure 1 — 34/c Mini RG-62 plenum	5779-34	NA	26 AWG Solid Silver Plated Copper	Foam FEP .104	T.C. Braid 90% coverage	Fluoropolymer clear 1.011	95	NA
Figure 2 — RG-6 type + 4-pair telephone composite plenum	0585-04	NA	A. 18" AWG Bare / B. 24 AWG Bare	A. Foam FEP .180 / B. Halar .031	A. Alum. Mylar + T.C. Braid / NA	A. FEP / NA — Overall Jacket FEP	75 / 100	50 / 1.5; 100 / 2.1; NA / NA
Figure 3 — 36/c composite	6459	NEC 725-38 (b) (3)	A. #24 AWG Solid Bare / B. #22 AWG Solid Bare / C. #22 Solid Bare	A. Halar[2] .034 / B. Halar[2] .040 / C. FEP[1] .068	A. No Shield / B. No Shield / C. Alum/Tape with #24 T.C. Drain Wire	Fluoropolymer Black .532		
Figure 4 — Voice, data, and video composite	6433	NEC 725-38 (b) (3)	C. #24 AWG Solid Bare Copper / D. #22 Solid Bare Copper / E. #22 Solid Bare Copper Covered Steel	C. Solid FEP[1] .037 / D. Solid FEP[1] .037 / E. Foam FEP[1] .103	C. Overall Tape and #24 AWG Solid Drain / D. Individual Alum/Tape and #24 AWG solid drain / E. T.C. Braid 95% Coverage	C. Fluoropolymer Clear .204 each pair / D. Fluoropolymer clear .226 each pair / E. Fluoropolymer .157		Overall Jacket Fluoropolymer .486"

1 = Trademark of de Nemours & Co.
2 = Trademark of Allied Corporation
Note: Please consult Graybar for designs not shown here.

Table 3-2 Composite building cables (Continued)

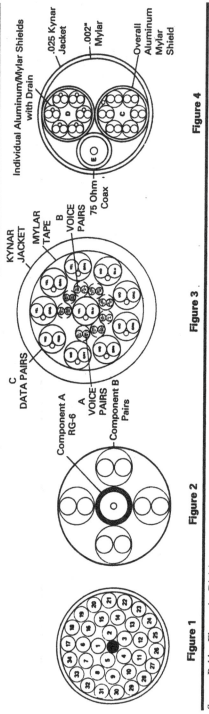

Individual Aluminum/Mylar Shields with Drain

.025 Kynar Jacket

.002" Mylar

Overall Aluminum Mylar Shield

75 Ohm Coax

Figure 4

KYNAR JACKET

MYLAR TAPE

B VOICE PAIRS

C DATA PAIRS

A VOICE PAIRS

Component A RG-6

Component B Pairs

Figure 3

Figure 2

Figure 1

Source: Belden Electronics Division

Figure 3-6 Coaxial cable connector: (a) BNC type, (b) TNC type, (c) plastic BNC type. (*Source:* TYCO Electronics Corporation)

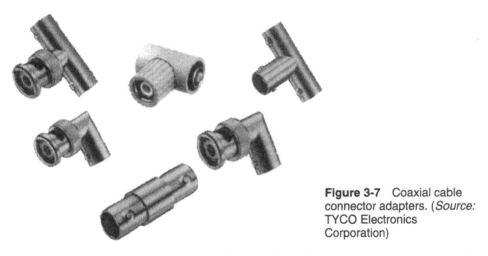

Figure 3-7 Coaxial cable connector adapters. (*Source:* TYCO Electronics Corporation)

Coaxial cable can be terminated with either a male or female connector, and various adapters can be connected to a coaxial cable to allow for parallel or series connections (Figure 3-7). However, there are several types of connectors that can be installed on coaxial cable. Figure 3-8 illustrates four types of TNC connectors. These 75-Ω impedance connectors have a frequency range of 2 Hz to 1 GHz.

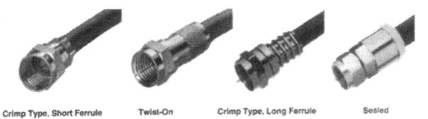

Crimp Type, Short Ferrule   Twist-On   Crimp Type, Long Ferrule   Sealed

Figure 3-8 TNC coaxial cable connectors. (*Source:* TYCO Electronics Corporation)

Figure 3-9 illustrates the installation method for crimp-type and solder-type BNC connectors. The *solder type* in Figure 3-9a requires soldering to the center probe that extends the conductor. Care must be take when assembling this connector to assure that the conductor is of the correct length, the pin is not overheated, and the ground braid (shield) is correctly installed in the holding shell and securing nut. The *crimp-type connector* (Figure 3-9b) is installed with a crimping tool and a wrench. The advantage of this type of connector is that no soldering is required, eliminating the possibility of overheating and melting the inside insulation between the center conductor and the ground shield (which could cause a short circuit as the connector is fitted).

Figure 3-10 depicts the screw-on coaxial connector. To attach this type of connector, the outer insulation, shield braid, and inner insulation are trimmed with a special coaxial cable stripper and the connector is screwed into the coax.

Table 3-3 summarizes the advantages of the two-piece connector, which requires only one crimping action. The detailed procedure for cable preparation with this tool is shown in Figure 3-11. This tool can also be used for preparing the cable for the other types of connectors. TNC-type cable requires that a crimper be used to secure the body of the connector to the shield. Figure 3-12 depicts such a crimper that also may be used for connecting an assortment of wire terminals.

When installing a workstation the connecting coaxial cable is run *home* to the source equipment (Figure 3-13a) or *daisy chained* with other devices within the computer/data network (Figure 3-13b). Regardless of the type of connection to the network, the cabling from the workstation will probably connect to a wall plate, such as shown in Figure 3-14. The workstations may be terminated from output to twisted-pair cabling, in which case the output impedance of the device would differ from that of the coax. Figure 3-15 shows an application that utilizes several different types of coaxial cable connectors.

When coax and twisted pair are connected, the different impedances must be matched by an impedance-matching device called a *balun*. The balun can be connected between the workstation and the face plate between the two different wiring media or at a patch panel. There is a multitude of choices for balun connections; Figure 3-16 depicts several types. Figure 3-17 shows the schematic diagram of a coax to twisted pair with an RJ11 connection.

3.4 GROUNDING OF COAXIAL CABLE

The necessity of grounding was discussed in Chapter 2, as well as the problems that can occur without proper grounding and the results of ground loops in the grounding of a cable. (A review of Section 1.12 on grounding and Section 2.4 on shielding may be in order.) In this section we will discuss only the grounding of coax as it pertains to the National Electrical Code (NEC). Figure 3-18 depicts the proper grounding techniques for coax that is run through metal conduit and into a service enclosure. The cable shield is grounded by a ground clamp to the bonding bushing of the conduit. The grounding conductor must be less than 15 m (50 ft) in length and AWG 2–8 solid copper; the insulation color should be green or green and yellow. Barrel connectors should

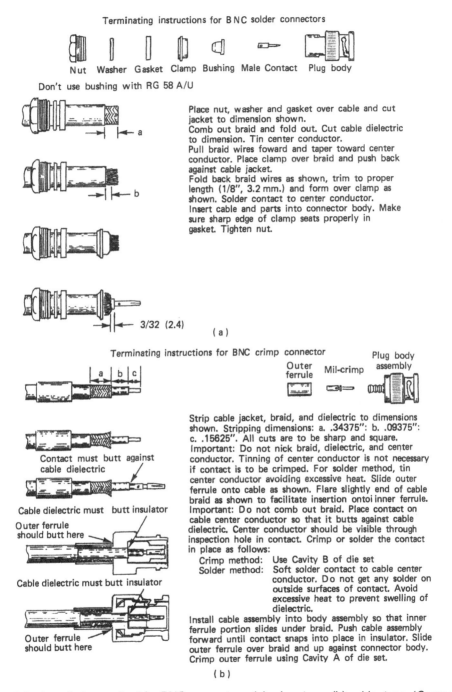

Terminating instructions for BNC solder connectors

Nut   Washer   Gasket   Clamp   Bushing   Male Contact   Plug body

Don't use bushing with RG 58 A/U

Place nut, washer and gasket over cable and cut jacket to dimension shown.
Comb out braid and fold out. Cut cable dielectric to dimension. Tin center conductor.
Pull braid wires foward and taper toward center conductor. Place clamp over braid and push back against cable jacket.
Fold back braid wires as shown, trim to proper length (1/8", 3.2 mm.) and form over clamp as shown. Solder contact to center conductor.
Insert cable and parts into connector body. Make sure sharp edge of clamp seats properly in gasket. Tighten nut.

3/32 (2.4)

(a)

Terminating instructions for BNC crimp connector

Outer ferrule   Mil-crimp   Plug body assembly

Contact must butt against cable dielectric

Cable dielectric must butt insulator

Outer ferrule should butt here

Cable dielectric must butt insulator

Outer ferrule should butt here

Strip cable jacket, braid, and dielectric to dimensions shown. Stripping dimensions: a. .34375": b. .09375": c. .15625". All cuts are to be sharp and square.
Important: Do not nick braid, dielectric, and center conductor. Tinning of center conductor is not necessary if contact is to be crimped. For solder method, tin center conductor avoiding excessive heat. Slide outer ferrule onto cable as shown. Flare slightly end of cable braid as shown to facilitate insertion onto inner ferrule.
Important: Do not comb out braid. Place contact on cable center conductor so that it butts against cable dielectric. Center conductor should be visible through inspection hole in contact. Crimp or solder the contact in place as follows:

Crimp method: Use Cavity B of die set
Solder method: Soft solder contact to cable center conductor. Do not get any solder on outside surfaces of contact. Avoid excessive heat to prevent swelling of dielectric.

Install cable assembly into body assembly so that inner ferrule portion slides under braid. Push cable assembly forward until contact snaps into place in insulator. Slide outer ferrule over braid and up against connector body. Crimp outer ferrule using Cavity A of die set.

(b)

Figure 3-9 Installation method for BNC connectors: (a) crimp type, (b) solder type. (*Source: Courtesy Anixter Corp.*)

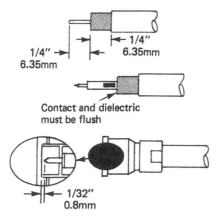

Contact and dielectric
must be flush

1. Using the Paladin ETN wire stripper or suitable alternate, prepare the end of the coaxial cable for connector instsllation.
2. Trim cable as shown, taking care not to nick the center conductor or outer braid.
3. Ensure that the outer braid lays flat.
4. Trim off any and all excess braid.
5. Twist the contact in a clockwise direction, on the inner conductor, untill the back end of the contact is flush with the inner dielectric.
6. Twist the connector onto the cable in a clockwise direction. The connector is properly installed when the end of the contact is positioned within 1/32" of the front edge of the connector.

Figure 3-10   Installation of a screw-on type of coaxial connector. (*Source:* TYCO Electronics Corporation)

Table 3-3   Advantages of the two-piece connector

- Two parts make up a plug or jack
- No danger of heat damage to coax
- Fully intermateable with comparable UG/U* series connectors
- Improved cable retention and insulation grip
- Ease of inspection
- Stabilized inner contacts
- Less critical stripping dimensions than required for solder assemblies
- Low VSWR†
- Reduced noise level
- Simplified replacement in field
- Full responsibility for development and performance crimp
- Positive insulation grip with a crimped braided ferrule
- Lightweight—¾ oz (23.33 g)(cable plug and jack)

* UG/U - underground utility cable

† VSWR - voltage standing wave ratio

be installed at the entrance of each building entry point, and all exposed connectors should be insulated to protect from the weather. Each entrance connector must be grounded.

The neutral bar on the inside of the service box is connected to the grounding bushing of the conduit and is, in turn, connected to a ground electrode. The grounding electrode conductor is typically of #2 to #8 AWG bare copper wire and must be run from the box through a cable bushing and conduit to the grounding electrode.

The conduit for the grounding conductor can usually be of electrical metal tubing (EMT) or plastic polyvinyl chloride (PVC). The grounding rod is a bronze rod driven at least 10 ft into

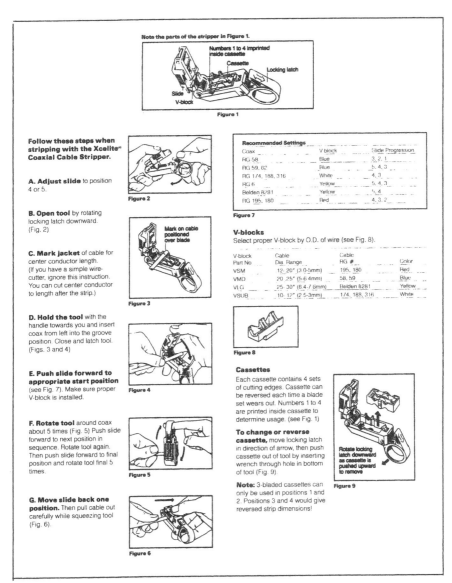

Figure 3-11 Coaxial wire stripper with details for operation. (*Source:* Courtesy Xcelite, Inc.)

the earth or grounding conductors embedded in the building foundation. For new facilities a grounding grid should be designed by the facility architect. Both local and national codes should be consulted to assure compliance.

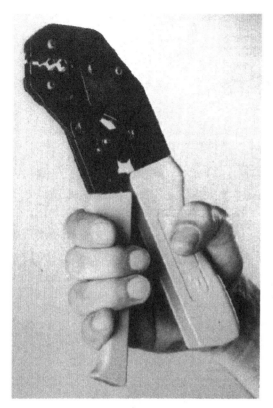

Figure 3-12   Mechanical crimping tool.
(*Source:* Courtesy AMP Inc.)

3.5 APPLICATION OF COAXIAL CABLE

Coaxial cable can be used for almost any wiring connections. Some examples of coaxial cable applications are as follows:

1. Closed circuit TV
2. Cable TV
3. Video security systems
4. Computer systems
5. Communication systems

Figure 3-19 illustrates an application of thin coax cable in an office environment.

3.6 ADVANTAGES OF COAXIAL CABLE

The following are the advantages of coaxial cable:

1. Low susceptibility to electromagnetic pick-up, resulting in less noise and cross talk
2. High bandwidth for transmitted signals, resulting in low signal distortion

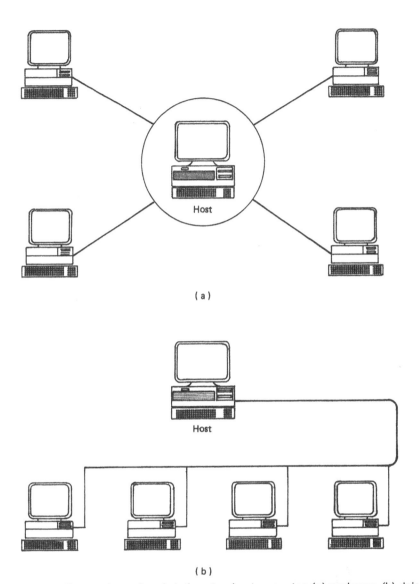

(a)

(b)

Figure 3-13 Connections of workstations to a host computer: (a) run home, (b) daisy chained to a patch panel (Figure 3-14) in a wiring closet or wiring.

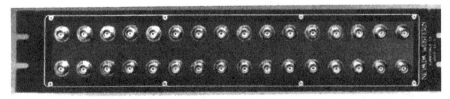

Figure 3-14 Patch panel. (*Source:* Courtesy Nevada Western)

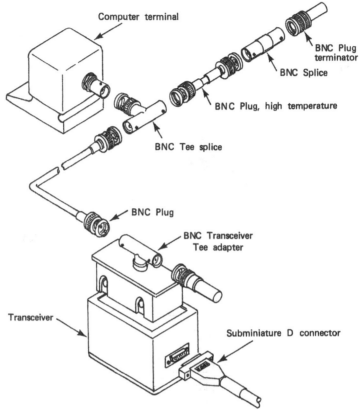

Figure 3-15 Application of coaxial connectors. (*Source:* TYCO Electronics Corporation)

Figure 3-16 Balun matching devices: (a) RJ-type connector to coax, (b) patch panel and breakout box. (*Source:* TYCO Electronics Corporation)

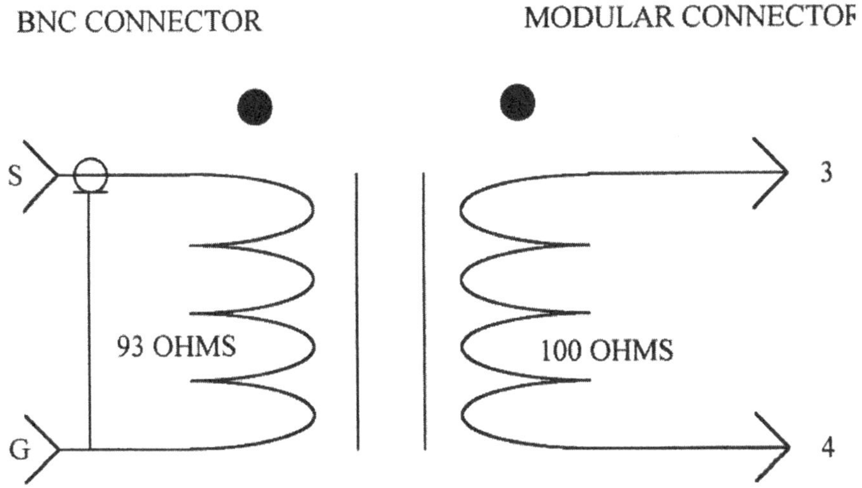

BNC CONNECTOR   MODULAR CONNECTOF

S

93 OHMS   100 OHMS

G

3

4

Figure 3-17   Schematic of a coax to twisted-pair balun.

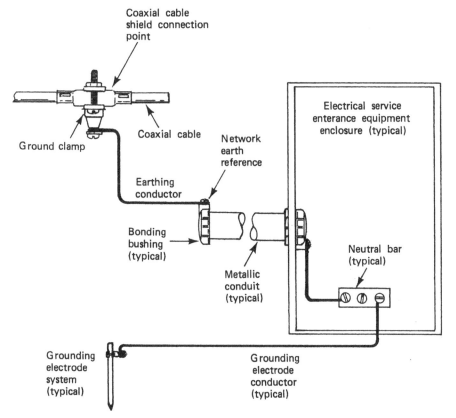

Figure 3-18   Proper grounding of coaxial entrance cable.

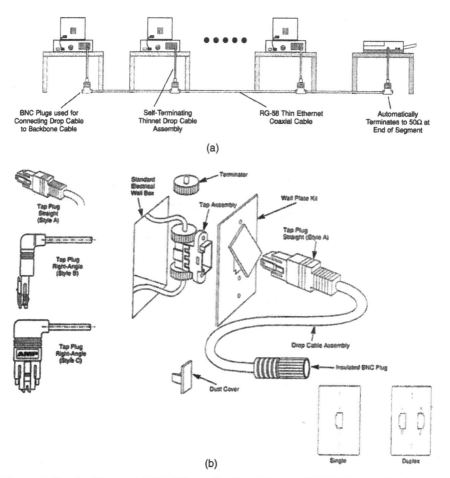

Figure 3-19 An Ethernet-10BASE2 application. (*Source:* TYCO Electronics Corporation)

3. Can be used for simultaneous transmission of voice, video, and data

4. Can be utilized over longer distances than twisted-pair cable

5. Proven performance and reliability over many years

6. Can be matched to operate with both fiber-optic and twisted-pair systems

7. A lower signal distortion due to phase shift, or variation of amplitude with increased frequency

8. Much larger number of channels can be transmitted over the same cable

9. Less cross talk between cables

10. Greater information security than twisted pair

3.7 DISADVANTAGES OF COAXIAL CABLE

The following are the advantages of coaxial cable:

1. More difficult to install than twisted pair
2. Heavier than twisted-pair or fiber-optic cables
3. Many systems have shifted from coaxial cable to twisted pair as new technology is developed to improve the transmission along twisted pair
4. Usually must be daisy chained or run home to work stations
5. Doesn't have the adaptability of twisted pair
6. Coaxial cable is more expensive and takes more time to install than twisted pair

3.8 SUMMARY

Coax is a time-proven cabling medium that can be utilized in most audio/data communication systems and with most telecommunication devices. The additional expense of coaxial cable is usually justified over twisted-pair cable if information security and signal bandwidth are important. Coaxial cable is less expensive to install than fiber-optic cable.

QUESTIONS

1. What are the advantages of coaxial cable over twisted pair?
2. When would the additional cost of coax over twisted pair be justified?
3. What is the purpose of a balun being connected between a workstation and a coaxial cable?
4. What is the purpose of placing a balun between a coaxial cable and a twisted-pair cable?
5. What is the major disadvantage of coax over twisted-pair wiring?

Fiber Optics

4.1 INTRODUCTION

The science of fiber optics deals with the transmission of energy in the form of light along a transparent medium such as glass or plastic. When one of these materials is formed into a transparent fiber and coated by a material that is less light conductive, an effective mirror is formed around the fiber. This mirror effect produces a **light pipe** through which light is transported. Light rays introduced into the fiber travel down the clear medium of the light pipe, reflecting back and forth from the edges of the fiber (Figure 4-1). The angle of reflection (leaving the mirror surface) is the same as the angle of refraction (entering the mirror surface). If the refraction angle is too great the light will not be reflected and will travel through the cladding much as light rays from the sun reflect or enter the surface of a lake at different times of day. The angle at

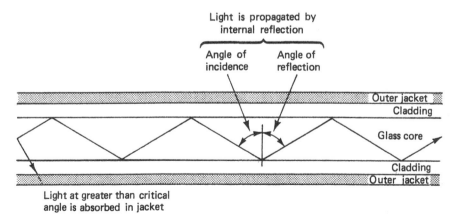

Figure 4-1 Light reflects back and forth from the walls of the fiber. (*Source:* Courtesy of Corning Cable Systems)

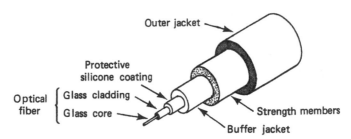

Figure 4-2 Cutaway view of a fiber-optic cable showing the fiber, the cladding, and the outer jacket.

which light rays will not be reflected from a surface is called the **critical angle**. We need not concern ourselves further with this phenomenon.

Fiber-optic cable is constructed with a small transparent center core that is covered by the cladding and sealed by an outer jacket. The cladding, usually of a less transparent glass than the fiber, protects the fiber and aids in the reflection process. A jacket of a tough plastic is formed over the cladding to protect both the cladding and the fiber from the outside environment, such as heat, moisture, dirt, scratches, and nicks. Figure 4-2 depicts a cutaway view of a fiber-optic cable showing the transparent core, the cladding, and the jacket.

Fiber optics, the technology for the transmission of the visible light portion of the electromagnetic frequency spectrum through a glass fiber, has been a technology waiting for widespread application. The need for wiring systems with greater signal capacity, greater bandwidth, and information security in the telecommunication industry has expanded the use of fiber-optic cables. Thousands of fiber cables are installed as overhead and underground links and service connections and as customer premise wiring.

Fiber cables, as compared with copper cables, have greater signal capacity, are unaffected by electromagnetic waves, create no electromagnetic radiation, and therefore offer extremely good information security. It is almost impossible to "steal information" from a fiber cable without being detected.

As we stated earlier, fiber-optic cables transmit signal information with light waves as the signal carrier and special glass or plastic fibers as the transmission medium. The transmission of light through a fiber is a rather complex science; a complete treatment is beyond the scope of this text and is not needed for installation and testing of fiber-optic cables. The characteristics of fiber light transmission and comparison of the different fiber types, regarding bandwidth and distance of transmission, will be discussed. This information should be helpful in selecting a fiber type for a given application and understanding installation and testing requirements.

4.2 FIBER TYPES

Optical fibers are classified by material composition, the refraction index of the core, and the modes (signal transmission methods) of the fiber. The **refraction index** is the angle that the light is reflected from the walls of the fiber. The **mode** is how the light waves are propagated (trans-

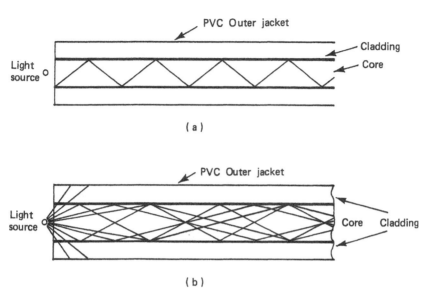

Figure 4-3   Light signal transmission along a fiber cable: (a) single-mode cable, (b) multimode cable.

mitted) through the fiber. Although usually glass, the fiber can be plastic. Light in the form of photons of energy is transmitted (propagated) down the fiber by what are called *modes*. Mode is a concept describing the way light waves travel through a medium. James Clark Maxwell, a nineteenth-century Scottish physicist, showed that electromagnetic waves are a single form of energy and that propagation follows strict rules, now known as Maxwell's law. For our purpose, mode is simply the path that light travels down a fiber (Figure 4-3). The number of modes that can be transmitted down a fiber cable range from one for a single-mode fiber cable to over a million for a multimode cable configuration.

The diameter of the fiber determines if a fiber is a single mode or multimode. When the fiber diameter is reduced to 5 μm, the fiber will support only one mode of light transmission. The standard cladding for both a single-mode and a multimode cable is 125 μm. This value was chosen because

1. It allows a size standardization.
2. The cladding must be approximately 10 or more times thicker than the core.
3. This value will make the fiber cable less brittle and easier to handle during installation.

There are two primary optical fiber attributes, bandwidth and attenuation, that must be considered before selecting a specific type of fiber for an application. **Fiber bandwidth** quali- fies the information-carrying capacity of a fiber. Bandwidth is measured in units of *Megahertz per kilometer* (MHz·km). The term refers to the bandwidth of the fiber and determines the max- imum cable length depending on the bit rate or protocol speed. As the data rate increases in giga-

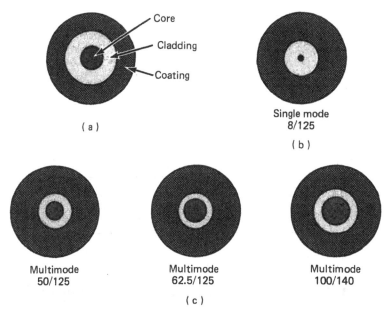

Figure 4-4   Types of fiber cable: (a) cross section of fiber cable, (b) single mode, (c) three sizes of multimode.

bits per second, the distance decreases. The advantages of greater bandwidth (MHz•km) are higher data rates for a given distance and longer cable connections at a given data rate.

Figure 4-4 depicts various types of fiber cables. Figure 4-4a shows a typical cross section of a fiber cable.

Single-mode fibers (Figure 4-4b) have a virtually unlimited bandwidth. Single-mode fibers are designed to carry only one mode of light and therefore do not experience modal dispersion like multimode fiber. The bandwidth of single-mode fiber is not limited by bandwidth but by attenuation, the electronic devices within the system, and system cost considerations. Single-mode fiber cable is best suited for long-distance connections, greater than 1 to 2 km. However, the cost of connections and transceivers is higher than for multimode fiber.

Multimode fiber cable (Figure 4-4c) is classified as multimode step-index fiber (step-index fiber) and multimode graded-index fiber. Multimode uses a parabolic (graded index) shape to minimize modal dispersion. This design maximizes bandwidth while maintaining low attenuation. The major restriction to network design for multimode fiber is bandwidth. Multimode fiber of 62.5/125 has a bandwidth in the 800 nm range of 160 MHz•km. Because of this limited range network designers are looking to a 50/125 multimode fiber that has been used in Japan for some time. The bandwidth of this fiber is 500 MHz•km, giving a 3× advantage over 62.5/125 fiber. The graphical analysis in Table 4-1 compares the bandwidth, data rate, and link length of the two fibers.

Table 4-1 Comparison of bandwidth, data rate, and link length of 50/125- and 62.5-µm fibers

Attribute	Units	50 µm	62.5 µm	SMF
Bandwidth - 850 nm	MHz•km	500	160	n/a
Bandwidth - 1300 nm	MHz•km	500	500	n/a
Attenuation - 850 nm	dB	2.5	3.0	n/a
Attenuation - 1300 nm	dB	0.8	0.7	0.4
Bending radius	mm	30	30	30
Relative system cost	%	100%	100%	400%

Source: Courtesy of Corning Cable Systems

Table 4-2 denotes the application standards for 50/125, 62.5/125 multimode fiber and single-mode fiber. The table compares data rates, wavelength, maximum link distance, and suggested protocol applications. Ten years ago 62.5/125 µm fiber was recommended as the industry standard. However, the short wavelength and high bandwidth requirements of present equipment have caused the industry to reevaluate this recommendation in favor of 50/125 µm fiber. At short wavelength 62.5/125 µm fiber has 50% of the bandwidth of 50/125 µm fiber.

4.3 LIGHT CONVERSION

Data, voice, and video are generated as electromagnetic energy by various types of transducers, such as microphones, scanners, video cameras, pulse generators, and speakers. The electromagnetic energy is converted to useful information by transducer devices, such as speakers, computers, cathode-ray tubes, and printers. Electromagnetic energy can be transported via wires or through space.

Fiber-optic cables act very effectively as pipes to transport voice, video, and data as light waves. Fiber-optic cables cannot transmit information as electromagnetic energy. Converters (transmitters) must be used at the send end of the fiber to convert the electromagnetic energy information into light transmissions, and another converter must be used at the receiving end of the fiber to convert the light energy to electromagnetic energy. A converter must also be used in repeaters.

Transmitters that use light-emitting diodes (LEDs) operating at data rates of up to 155 Mbps at either short wavelengths (850 mm) or long wavelengths (1300 mm) have been commonly used for transmitting information over multimode optical fiber. LEDs used in telecommunications emit light in the near infrared range and are limited to maximum data rates of 622 Mbps. LEDs emit light in a broad cone that can only be captured by the large numerical aperture of multimode fiber.

Table 4-2 Applications and standards of fiber-optic cables

Application	Data Rate (Mb/sec)	Wavelength (nm)	BW (MHz•km)		Max. Length (meters)		
			50 μm	62.5 μm	50 μm	62.5 μm	SMF
Ethernet - 10BaseF	10	850	500*	160	1000*	2000	—
Ethernet - 100Base F	100	1300	500*	500	2000	2000	2000
Ethernet - 1000Base-SX	1000	850	500*	160†	550	220	—
Ethernet - 1000Base-LX	1000	1300	500†	500	550	550	5000
Token Ring	16	850	500*	160	1000*	2000	—
FDDI PMD	100	1300 MMF	500*	500	2000*	2000	—
FDDI-LCF	100	1300 MMF	500*	500	500*	500	—
FDDI-SMF	100	1300 SMF	—	—	—	—	58000
Fibre Channel	1063	780/850	500	160	500	175	—
Fibre Channel	531	780/850	500	160	1500	350	—
Fibre Channel	531 & 1063	1300 SMF	—	—	—	—	10000
ATM	155	780/850	500‡	160	2000‡	2000	—
ATM	155	1300	500	500	2000	2000	55000
ATM	622	1300 LED	500	500	500	500	—
ATM	622	780/850	500‡	160	500‡	500	—
ATM	622	1300 SMF	—	—	—	—	15000

* Addressed in informative annex of the standard

† Additional specified bandwidth options are 50 μm 400/400 MHz•km and 62.5 μm 200/500MHz•km

‡ Using an uncommon 50 mm BW of 160 MHz•km, the link lengths would be 1,000 and 300 meters for 155 and 622 Mb/s respectively

Source: Courtesy of Corning Cable Systems

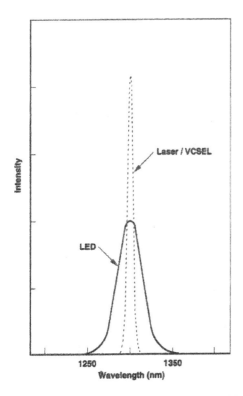

Figure 4-5 Pulse width of light source. (*Source:* Courtesy of Corning Cable Systems)

Lasers emit light in a narrow beam, making them ideal for use with the small numerical aperture of single-mode fiber. However, lasers are also used with multimode fiber in high-bit-rate systems such as Gigabit Ethernet and 2.5-Gbps asynchronous transfer mode (ATM). To achieve faster data rates such as Gigabit Ethernet, 2.5 Gbps ATM and beyond, *vertical cavity surface emitting lasers* (VCSELs) are used. Lasers are relatively inexpensive when their reliability and their high bit rate is considered.

Fiber-optic transmitters are classified by the wavelength at which they emit light. The nominal emission is the center wavelength, λ_c, of the transmitter. The transmitted signal is actually a collection of wavelengths around a center frequency. The nominal frequency is a function of the materials used in the manufacture of laser transmitters. Figure 4-5 compares the bandwidth of LEDs and lasers in wavelength versus intensity of light transmission. The total power produced by an optical transmitter is spread over a spectrum around the center wavelength. The range is specified by the spectral width (Figure 4-6), $\Delta\lambda$, measured in nanometers and affects the overall transmission capacity of the fiber-optic link. $\Delta\lambda$ is usually expressed as a *full-width half-maximum* (FWHM) value, as a *root-means-square* (RMS) value, or as a –3 dB value. Spectral widths vary from a few nanometers for light emitting diodes (LEDs) and vertical cavity surface emitting diodes (VCSELs) to over 100 nm for LEDs.

Receivers like transmitters are required in each piece of fiber-optic transmission equipment. Optical receivers incorporate a photodetector such as a photodiode to covert the incoming

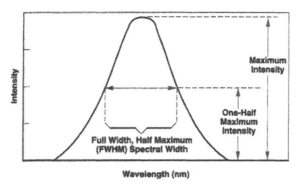

Figure 4-6 Spectral comparison of LEDs to lasers.
(*Source:* Courtesy of Corning Cable Systems)

optical signal back to an electrical signal. The operating wavelength of the photodiode must match that of the transmitter. A photo receiver designed for 850 nm will probably not function efficiently at 13 nm.

4.4 CABLE CONSTRUCTION

Fiber-optic cables are constructed in special forms to serve the application in which they are placed. However, all are comprised of a fiber light conductor, a light-reflecting cladding, and an outer jacket. The cladding protects the fiber, acts as a light reflector (to prevent the light below the critical frequency from escaping from the fiber), and seals the light in the cable.

The function of the outer coating is to protect the cladding and the fiber. The outer coating, comprised of one or more layers of polymer (plastic), protects the interior from moisture and physical abuse, such as dirt and nicks.

Figure 4-7 is a summary of several packing techniques for fiber cable. Figure 4-7a shows a single fiber called simplex fiber. While a single fiber will carry thousands of separate data streams, its disadvantage is that a break would put the entire system out of service. Figure 4-7b depicts a zip cable with two separate fibers. This cable has twice the capacity of the simplex cable and the advantage of a back-up cable. The zip cable is a dual cable that can easily be separated. It is flatter than the other type of dual cable. Figure 4-7c shows a dual fiber in one cable package with a strong center strength member that protects the fiber from stress and strain. Figure 4-7d is a six-fiber cable. This cable probably has more conductors than necessary for most applications. However, the designer should plan on an order of three times the base requirements for expansion, and each fiber has individual security. A strength member can be included in any type of cable and should always be included in a cable that is to be pulled through a conduit or duct or hung between buildings.

Technically speaking, the entire make-up of fiber and covering is referred to as a cable. However, here we will refer to all the outer covering that protects one or more internal fibers from the stress and strain of the external environment as cabling. A buffered jacket, which protects the fiber and the cladding, is put on the cable by the fiber manufacturer. Additional buffering mate-

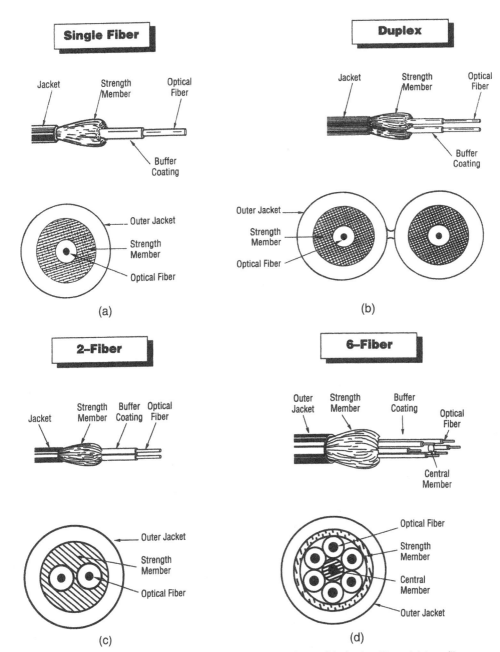

Figure 4-7 Packing methods for fiber cable: (a) single fiber, (b) duplex fiber, (c) two-fiber, (d) six-fiber. (*Source:* Courtesy of Belden, Inc.)

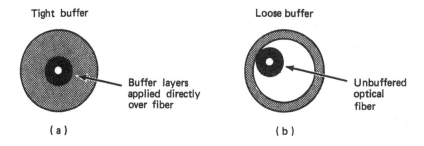

Figure 4-8 Types of buffering: (a) tight buffered cable, (b) loose buffered cable, (c) comparison of the characteristics of the two types.

rial, the strength member, and the outer jacket are usually put on the cable by the cable manufacturer. These materials are chosen to protect the cable in a particular environment and to color code the cable. Two techniques are used for this final buffering: loose buffering and tight buffering. When tight buffering is utilized the buffering is applied tightly over the cladding and the fiber. This forms a solid tough cable that is flexible and allows tight radii in bends with low losses. Tight buffering has the additional advantage of high impact and crush resistance. However, tight buffering results in less isolation of the fiber from temperature variations (Figure 4-8a).

Loose buffering (Figure 4-8b) is formed by adding a loose covering, much like a sleeve, over the cladding. The covering is usually much larger than the fiber and is gel filled. The gel and buffering tube isolate the fiber from external mechanical stress and heat. The degree of fiber shrinkage can be controlled, thereby reducing attenuation due to temperature. Another advantage of loose buffering is that the buffer tube is designed to take any stress or strain placed on the cable and relieve the fiber from change of length due to retraction or expansion with temperature change. The buffer tube can carry more than one fiber cable.

Each buffering type offers advantages. Tight buffering permits small, lighter cable design, and a more flexible, more crush-resistant cable. A summary of the advantages of each type of buffering is shown in Table 4-3.

Strength members may be made of steel, fiberglass, or Aramid yarn. Steel, while the strongest, should not be used in all optical cables. However, it offers a very strong support for cables

Table 4-3 Loose and tight buffering comparisons

| | CABLE STRUCTURE | | |
CABLE PARAMETERS	Loose Tube	Tight Buffer	Breakout
Bend radius	Larger	Smaller	Larger
Diameter	Larger	Smaller	Larger
Tensile strength	Higher	Lower	Higher
Impact resistance	Lower	Higher	Higher
Crush resistance	Lower	Higher	Higher
Attenuation change at low temperature	Lower	Lower	Higher

that must be strung between buildings or poles. Cable loads during installation may place a tensile stress on the fiber, causing an attenuation increase and possible fiber fatigue. Optical fibers stretch very little before separating so that strength members must have low elongation at the expected stress loads. Table 4-4 compares the tensile properties of strength members.

Table 4-4 Strength member comparison

	Load to Break (lb)	Diameter (in)	Elongation Break (%)	Weight (lbs/Kft)
Steel	480	0.062	0.7	7.5
Aramid	944	0.093	2.4	1.8
Epoxy fiberglass rod	480	0.045	3.5	1.4

Source: Courtesy of Corning Cable Systems.

The small size of fiber and cladding allows cables to be formed in almost any special configuration. For example, the cable under carpet in Figure 4-9a would hardly make a ripple under a carpet because it is only 0.07 in thick. Another unique application is thin ribbon cable (Figure 4-9b), which has an outer coating that allows great flexibility.

4.5 CABLE CHARACTERISTICS

As stated earlier, *dispersion* is the spreading of a light pulse as it travels along optical cable. Dispersion limits the bandwidth and information-carrying capabilities of a cable. The bit rate can be

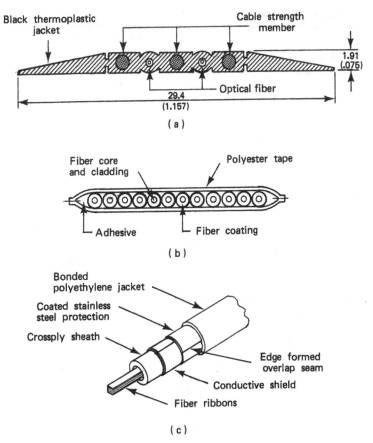

Figure 4-9 Special cable configurations for fiber-optic cables due to their small size: (a) under-carpet cable, (b) ribbon cable, (c) cable construction.

high enough that the pulses overlap and become unintelligible. Figure 4-10 illustrates the problem of pulse spreading.

Dispersion or spreading for step-indexed fiber is 15 to 30 ns•km. This means that two modes entering a fiber can be separated by 30 ns in a 1000-m cable. Eventually, two pulses could combine to form an unintelligible pulse. Graded-index fiber has a core of numerous concentric layers of glass. Each layer of the core refracts light. Instead of being sharply refracted, as in step-indexed fiber, the light is bent in an almost sinusoidal pattern. The light near the center has the

Figure 4-10 Pulse spreading.

lowest velocity and the light near the outer rings has the greatest velocity. This produces a more even travel rate and reduces the pulse-spreading effect.

The bit rate of the pulses must be low enough as not to overlap due to dispersion. There are three main types of dispersion in fibers: modal, material, and waveguide. However, we will not concern ourselves with the specific details of each of these since cable manufacturers do not differentiate among the types.

Cable manufacturers do not specify a dispersion factor for multimode cables. Instead a *figure of merit* called the *bandwidth-length product* or simply bandwidth is given in megahertz per kilometer. A figure of 500 MHz•km means that a 500-MHz signal can be reproduced after a transmission of 1 km, or a 250-MHz signal can be reproduced after a transmission of 2 km. For example,

1000 MHz•0.5 km
500 MHz•1 km
250 MHz•2 km
100 MHz•5 km

The dispersion factor for single-mode cable is expressed in picosecond per kilometer per nanosecond (p•km•ns) of source spectral width. This means that in single-mode cable the dispersion factor is most affected by the source spectrum (frequency or light). The wider the source spectrum or the greater the number of wavelengths of the source light, the greater the attenuation.

Signal attenuation results in a loss of signal power. Attenuation in a fiber-optic cable is the loss of light as the signal travels along the cable. Attenuation varies as the wavelength or frequency of the light varies. Some wavelengths or frequencies travel with little or no loss while others have a large loss. Figure 4-11 shows a chart of decibel attenuation versus the wavelength of light over the wavelength of 700 to 1600 mm for a multimode cable. Figure 4-12 shows the attenuation characteristics of a single-mode cable over an approximate wavelength. Note that a frequency of 200,000 MHz has a wavelength of 1.5 μm. The wavelength of a waveform is the distance that the wave travels in the time that it takes to complete one cycle. A wave travels approximately 186,000 miles in one second. In the formula below wavelength in meters is indicated by the Greek letter λ. The speed of light is represented by the letter C, and frequency is represented by the letter f.

$$\lambda = C/f$$

where
λ = wavelength in meters
C = speed of light (186,000 miles•s)
f = frequency in cycles per second (Hz)

$$\lambda = 300/f \text{ m}$$
$$= 300/2 \times 10^{11} = 1.5 \text{ nm}$$

Characterization data

Characterized parameters are typical values.

Core material index of refraction (peak):   1.4805 at 850 nm
                                              1.4748 at 1300 nm

Spectral attenuation (typical fiber):

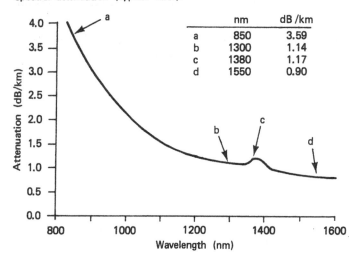

	nm	dB /km
a	850	3.59
b	1300	1.14
c	1380	1.17
d	1550	0.90

Figure 4-11 Attenuation in decibels per kilometer versus wavelength over 0.8 to 1.6 μm. (*Source:* Courtesy of Corning Cable Systems)

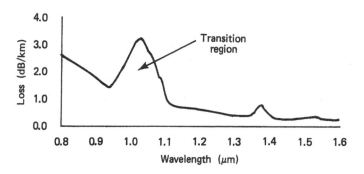

Figure 4-12 Attenuation in decibels per kilometer versus wavelength for a single-mode cable. (*Source:* Courtesy of Corning Cable Systems)

   When long cable runs are necessary, we can study the manufacturer's frequency versus attenuation charts and avoid the high-loss frequency areas. Of all the advantages of fiber optic cables over copper cables, the most important is that within the bandwidth of a signal mode, the loss in the fiber cable is constant.

4.6 DATA RATE

The data rate of a device is the maximum number of bits that can be transmitted in one second and received with a *bit error rate* (BER) below a certain level. A typical BER is one error in one Giga (10^9) bits to one Tera (10^{12}) pulses (bits).

One of the limiting factors of any digital device is the time for the device to turn on (rise time) and turn off (fall time). In the case of light transmission this is the time for the light to turn on and the time for the light to turn off. The rise time is the time the transmitted pulse takes to rise from 10 to 90% of its value. The fall time is the time that a transmitted pulse takes to fall from 90 to 10% of its value.

4.7 SPECIFYING A TYPE OF FIBER-OPTIC CABLE

When specifying a data, voice, or video communications network, optical fiber should be considered. When the fiber is selected as a communication medium, the network designer must select the appropriate fiber type that meets the system proposed bandwidth and standards. The worst-case bandwidth is determined by the worst-case protocol that will accomplish the current and future needs of the organization. Table 4-5 denotes Gigabit Ethernet standard. For example, if the worst case is Ethernet 1000Base-SX or LX at a link distance of 500 m, then 50/125 fiber would be the choice for SX cable, and both 20/125 and 62.5 fiber could be used for LX application. This can be validated by observing that both 50 m and 62.5 m have a bandwidth of 500 MHz•km that satisfies the 400 MHz•km requirement of 1000Base-LX. On the other hand, 50 μm has a 500 MHz•km bandwidth in SX band.

Table 4-5 compares 50/125 μm fiber with 62.5/125 μm fiber cable. The table compares the attenuation and bandwidth of the two types of cable. Regardless of the fiber type chosen, the designer should specify the attenuation and bandwidth for multimode fiber at both 850-nm and 1300-nm windows, and single-mode fiber at both 1310-nm and 1550-nm windows. Table 4-6 lists the 50-μm standards.

Table 4-5   Comparison of the characteristics of 50-μm and 62.5-μm fiber cable (*Source:* Courtesy of Corning Cable Systems)

Standard	Fiber Type	Diameter (microns)	Modal Bandwidth (MHz·km)	Minimum Range (meters)
1000-SX (850 nm)	MM	62.5	160	220*
"	MM	62.5	200	275*
"	MM	50	400	500
"	MM	50	500	550**
100-LX (1300 nm)	MM	62.5	500	550
"	MM	50	400	550
"	MM	50	500	550
"	SM	9	NA	5000

Notes: *The TIA 568 building wiring standard specifies 160/500 MHz•km multimode fiber.
        **The ANSI Fibre Channel specification specifies 500/500 MHz•km 50 micron multimode fiber and
          500/500 MHz•km fiber has been proposed for addition to ISO/IEC 11801.

Table 4-6   Standards for 50-μm fiber cable (*Source:* Courtesy of Corning Cable Systems)

Fiber Standards:	
ANSI/TIA 492AAAB	Detail Specification for 50 μm Multimode Fiber (N. American standard)
IEC 793-2:1992	Product Specification for 50 μm Multimode Fiber (Int'l standard)
Cabling Standards:	
ISO/IEC 11801	Generic Cabling for Customer Premises (Int'l standard)
ANSI/TIA/EIA 568-A	Commercial Building Telecommunications Cabling Standard
	(N. American standard - accepted by TR-41.8 working group)
Applications Standards:	
IEEE 802.3	Ethernet*, Fast Ethernet*, Gigabit Ethernet
NCITS T-11	Fibre Channel
PHY Group	ATM Forum
ANSI X3T9.5	FDDI*
IEEE 802.5	Token Ring*
	*Addressed in informative annex of standard

The final step in installing the cabling system is to prepare a *Request for quote* (RFQ) from fiber suppliers and/or cable installers to elect bid proposals and ensure that the organization will get what it pays for.

The remainder of this chapter is for the cable installer or communication specialist who will install, test, certify, and maintain the fiber cabling.

4.8 INSTALLATION CONSIDERATIONS

The installation of twisted-pair or coaxial cable requires care to prevent broken or shorted conductors and strain that might cause future shorts. Fiber-optic cable is relatively fragile compared with twisted pair and coax with regard to tensile strength and bend allowance. Some of the precautions for installing fiber are discussed in this section.

4.8.1   Bending Capacity

Optical fiber is very flexible and capable of relatively tight bending. TIA[*] 568-A allows for a 30-mm (1.18-in) bend radius for two- and four-fiber cable. Since all premise fiber has the same size cladding, they all have the same flexibility. Signal attenuation and bandwidth are affected with only small bend diameters of less than 20 mm (1 in). Table 4-7 lists the bend radius recommendations for various fiber cables. Bending the cable tighter than the recommendation may result in attenuation and/or broken fiber.

* Telecommunication Industry Association.

Table 4-7   Bend radius specifications
(*Source:* Courtesy of Corning Cable Systems)

| Application | Fiber Count | Minimum Bend Radius | | | |
| | | Loaded | | Unloaded | |
		cm	in	cm	in
Campus	2-84	22.5	8.9	15.0	5.9
Backbone	86-216	25.0	9.9	20.0	7.9
Building	2-12	10.5	4.1	7.0	2.8
Backbone	14-24	15.9	6.3	10.6	4.2
	26-48	26.7	10.5	17.8	7.0
	48-72	30.4	12.0	20.3	8.0
	74-216	29.4	11.6	19.6	7.7
Horizontal	2	6.6	2.6	4.4	1.7
Cabling	4	7.2	2.8	4.8	1.9

*(1) Specifications are based on representative cables for applications.
    Consult manufacturer for specifications of specific cables.
* For open-systems furniture applications where cables are placed,
    Siecor's 2- and 4-fiber horizontal cables have a one-inch bend radius.
    Please contact your Siecor representative for further information.*

4.8.2   Tensile Rating

The tensile rating of fiber cable is indicated by the cable manufacturer and must not be exceeded during or after installation. Tension on the cable, other than hand pulls, should be monitored when the cable is pulled. Circuitous pulls can be achieved through center pull techniques or backfeeding. Pull boxes should be supplied for every third 90° bend to prevent binding of the cable. No residual tension should remain on any type of cable after installation other than that due to the cable's own weight. Table 4-8 identifies tensile ratings for fiber cables.

4.8.3   Vertical Rise of Fiber Cable

The vertical rise of any cable is restricted due to tension on the cable caused by its own weight and tensile strength. Vertical rise limits represent the maximum distance the cable can be installed without intermediate support points between the primary supports. Guidelines for vertical installation of fiber cable are as follows:

- All vertical cables must be secured at the top of run.
- Attachment points should be selected in accord with the vertical rise limits and the bend allowance of the cable.
- Long vertical cables should be secured with intermediate supports when the vertical rise limits have been reached.

Table 4-9 specifies typical vertical rise distance values.

Table 4-8 Typical Tensile Rating of Fiber Cable
(*Source:* Courtesy of Corning Cable Systems)

Application	Fiber Count	Maximum Tensile Load			
		Short Term		Long Term	
		N	lbs	N	lbs
Campus Backbone	2-84	2700	608	600	135
	86-216	2700	608	600	135
Building Backbone	2-12	1800	404	600	135
	14-24	2700	608	1000	225
	26-48	5000	1124	2500	562
	48-72	5500	1236	3000	674
	74-216	2700	600	600	135
Horizontal Cabling	2	750	169	200	45
	4	1100	247	440	99

*(1) Specifications are based on representative cables for applications.
Consult manufacturer for specific information.*

Table 4-9 Typical Vertical Rise Values for Fiber Cable
(*Source:* Courtesy of Corning Cable Systems)

Application	Fiber Count	Maximum Vertical Rise	
		m	ft
Campus Backbone	2-84	194	640
	86-216	69	226
Building Backbone	2-12	500	1640
	14-24	500	1640
	26-48	500	1640
	48-72	500	1640
	74-216	69	226
Horizontal Cabling	2	500	1640
	4	500	1640

*(1) Specifications are based on representative cables for applications.
Consult manufacturer for specifications of specific cables.*

4.8.4   Pathways and Spaces

The TIA/EIA[†] *Commercial Building Standards for Telecommunications Pathways and Spaces* addresses the requirements of pathways and spaces for cabling with TIA/EIA-569. The *telecommunications distribution methods manual* for the Building Industry Consulting Service Interna-

† Electronic Industry Association.

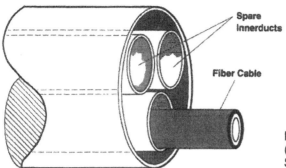

Figure 4-13 Duct and inner ducts. (*Source:* Courtesy of Corning Cable Systems)

tional (BICSI) is another resource for the network designer or wiring contractor that addresses this question.

4.8.5   Cable Protection

All cable should be installed in conduit when it is in an area that is exposed to possible damage from any type of stress. When cable is enclosed in a conduit, the installer must consider the possibility that additional cable will have to be pulled in the future. When a cable is pulled against other cables in a conduit, there is a possibility that the insulation of the cables will bind and cause stress or breakage, even with the use of a pulling lubricant. A solution is the addition of one or more inner ducts as the conduit is laid. Figure 4-13 depicts a conduit with inner ducts or conduits.

4.8.6   Fill Ratio

When a network is designed the designer should consider that the network will probably be obsolete in the future as new devices are installed, additional employees are added, and additional space is required. The addition of extra fibers or conductors in a conduit or duct should be considered. Extra wires or fibers installed can be very cost effective in the future. When pulling cable through conduit or ducts, less than 50% fill by cross-sectional area is recommended. Figure 4-14 illustrates this rule.

4.8.7   Building and Fire Codes

The National Electrical Code (NEC), while advisory, is the industrial standard for fire standards for all wiring in the United States. Chapter 5 covers telecommunication wiring fire codes.

4.8.8   Termination Spaces

Optical fiber hardware is typically designed for rack and/or wall mount. Space requirements for patch panels and jumper routing to electronics must be considered. Guidance is provided in the TIA/EIA *Commercial Building Standards for Telecommunications Pathways and Spaces* and the BICSI *Telecommunication Distribution Hardware and Outlets.*

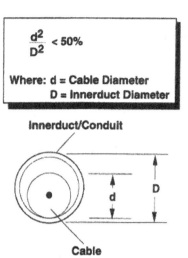

Figure 4-14   Fill ratio for conduit.

4.8.9   Aerial Cable

Aerial cable may be exposed to stress from wind or ice loading and to extreme temperature change. Aerial cable can be damaged by lightning, rodents, gunfire, and vehicular accidents. There is no safety net for some of these hazards. However, there are some guidelines to follow in the installation of outdoor cable:

- Pole spans of less than 300 ft can be engineered when a dedicated messenger wire strength member is used. Fiber manufacturers should be consulted when spans of over 300 ft are required.
- Cable can sometimes be overlashed or hung from existing cable. Care must be taken not to overload the existing cabling.
- Tight buffered fiber cable is not recommended for overhead cable, as any stress on the buffering reflects directly on the fiber.

4.8.10   Direct Burial Cable

While direct burial cable experiences fewer environmental hazards than aerial cable, it has its enemies. Primary dangers to direct burial cable are trenching and rodent damage. Rodent protection is provided with the use of steel tape armoring. Installation in ducts or conduits protects from rodents and offers some protection from dig-ups. Damage to burial cable can be reduced by taking certain precautions during installation:

- Cable should be buried 30 in deep.
- Cable should be buried below the frost line to prevent damage from freezing.
- Direct buried routes should avoid underground utilities.
- Plastic conduit of 1.5-in diameter will limit rodent damage.

- During burial a means of locating the cable should be included, especially for all dielectric fiber cable.
- Tight buffered cable is not recommended for direct burial as it is not waterproof.

4.9 CABLE TERMINATION

One of the final jobs in the installation of fiber cable is termination. There are four basic ways to terminate optical fiber:

1. Preconnectorized cable assemblies
2. Pigtail splicing
3. Field connectorization
4. Preconnectorized hardware

These options are available for both backbone cables and horizontal distribution cables. Each option has advantages and disadvantages. The selection depends on the installation and the experience of the installer.

Any restriction to the passage of light through a fiber increases signal attenuation. For this reason special care must be taken when attaching connectors or splicing the cable. Splicing is usually required in very long cables because cable is easier to purchase in the 1 to 5 km length. It is also very difficult to pull a cable of, say, 20 km.

Splicing and connectors are necessary when adding new cable to an existing system, connecting exterior to interior cable, splitting cable information into several work positions, connecting cable to a transmitter or receiver, breaking the system into several subsystems, and connecting the system into a switching system. Forming the fiber into several subsystems allows for ease of service, changing equipment as personnel change, and upgrading equipment without disturbing the system. There are many manufacturers producing a number of different types of connectors. Figure 4-15 depicts three industry standards. Adapters can be purchased to interconnect any of these connector types. The biconic type is recommended for single mode and the ST type is recommended for multimode operation. All these connectors can be used with adapters, which can be purchased to adapt one type of connector to the other. The biconic-type connector is used most often for single-mode operation and the ST[‡] type is most often used for multimode application. Figure 4-16 depicts several P/N type of connectors.

The features of an ideal connector are that it is economical, has low loss, is easily installed, and has reliability. A connector must be economical to install. This means that installation can be accomplished quickly and reliably by employees with a minimum of training. A connector should present a minimum attenuation (dB) loss to the signal. A connector should perform reliably and consistently after many connects and disconnects. The initial cost is probably the least important consideration for purchasing a given type of connector. Losses in an

‡ Trademark of American Telephone and Telegraph Company.

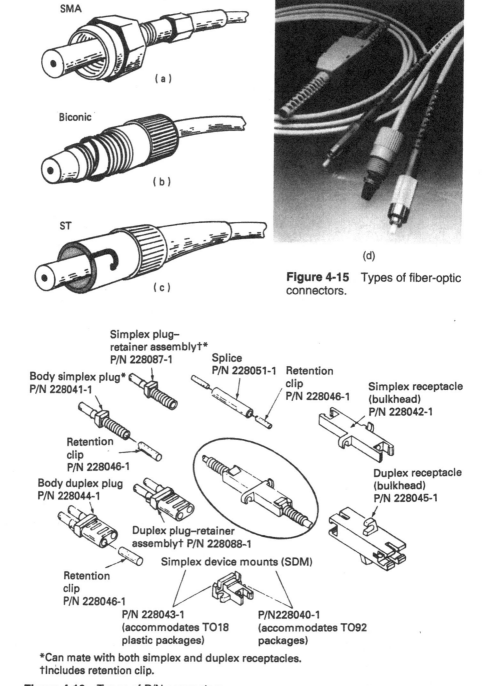

SMA

(a)

Biconic

(b)

ST

(c)

(d)

Figure 4-15 Types of fiber-optic connectors.

Simplex plug–
retainer assembly†*
P/N 228087-1

Splice
P/N 228051-1   Retention
clip
P/N 228046-1

Body simplex plug*
P/N 228041-1

Simplex receptacle
(bulkhead)
P/N 228042-1

Retention
clip
P/N 228046-1

Body duplex plug
P/N 228044-1

Duplex receptacle
(bulkhead)
P/N 228045-1

Duplex plug–retainer
assembly† P/N 228088-1

Retention
clip
P/N 228046-1

Simplex device mounts (SDM)

P/N 228043-1
(accommodates TO18
plastic packages)

P/N228040-1
(accommodates TO92
packages)

*Can mate with both simplex and duplex receptacles.
†Includes retention clip.

Figure 4-16 Types of P/N connectors.

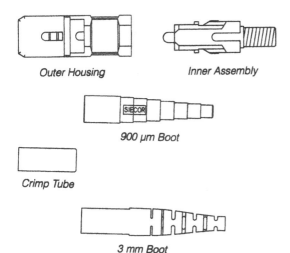

Outer Housing   Inner Assembly

900 µm Boot

Crimp Tube

3 mm Boot

Figure 4-17   Exploded view of an SC-type connector. (*Source:* Courtesy of Corning Cable Systems)

interconnection are caused by imperfections within the fiber itself, losses due to the connection or installation of the connector, and factors that relate to the system, such as matching of the cable to equipment. Figure 4-17 shows an exploded view of an SC-type connector.

The type of connector called "quick mount" is the answer for easy-to-install, economical connectors that can be installed quickly in the field. These types of connectors reduce labor cost and training time. Type UV-curable or glass-insert connectors feature a glass insert surrounded by ceramic (Figure 4-18). The glass tube propagates light so that a UV-curable adhesive can be used to bond the fiber to the ferrule. The bonding process takes about 45 seconds. The glass insert protrudes beyond the ceramic sleeve and the fiber protrudes beyond the glass insert. The insert and fiber are of approximately the same hardness, allowing the two to be polished at the same rate. The result is a flat point of contact surface that typically yields a 99% or greater success rate in the field. UV-cured connectors can be used for all types of connections, such as main, intermediate, and horizontal cross connects as well as multiple user outlets. Figure 4-18 is a cross-sectional view of a glass-insert connector before polishing.

The UniCam™ connector, depicted in Figure 4-19, is a mini-pigtail housed in a connector body with a fiber stub bonded into the ferrule and the endface polished at the factory to a physical contact (PC) finish. The other end of the fiber is cleaved and resides inside the connector. The field fiber is cleaved and inserted into the connector butting against the fiber stub. A rotating cam actuation process completes installation; no polishing or epoxy is required. The advantages

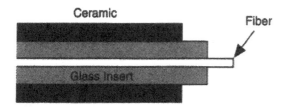

Ceramic   Fiber

Glass Insert

Figure 4-18   Glass-insert connector. (*Source:* Courtesy of Corning Cable Systems)

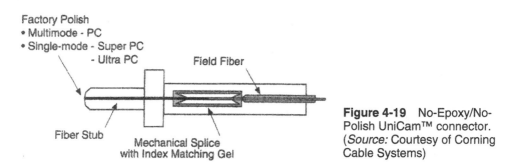

Factory Polish
• Multimode - PC
• Single-mode - Super PC
                    - Ultra PC

Field Fiber

Fiber Stub

Mechanical Splice
with Index Matching Gel

Figure 4-19 No-Epoxy/No-Polish UniCam™ connector. (*Source:* Courtesy of Corning Cable Systems)

of the no-cure UniCam connectors are ease of installation, full protection from the environment, long life, and minimum tool requirements. Installation of this type of connector requires only a stripper, a cleaver, a workstation tool, and an alcohol pad. Assembly is simple and easy with no consumables.

When a fiber is spliced the splice should be protected from strain or contamination. This can be accomplished by placing the splice in a splicing tray such as shown in Figure 4-20. The tray allows a recommended bend radius of the fiber and protects the splice.

Figure 4-21 shows the breakout of a multifiber cable into individual fiber elements. This method separates and protects the fiber elements so that they can be routed to individual equip-

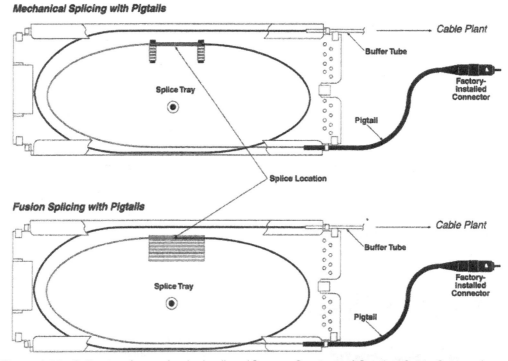

Mechanical Splicing with Pigtails

Cable Plant

Buffer Tube

Factory-installed Connector

Splice Tray

Pigtail

Splice Location

Fusion Splicing with Pigtails

Cable Plant

Buffer Tube

Factory-installed Connector

Splice Tray

Pigtail

Figure 4-20 Splice tray for mechanical splice. (*Source:* Courtesy of Corning Cable Systems)

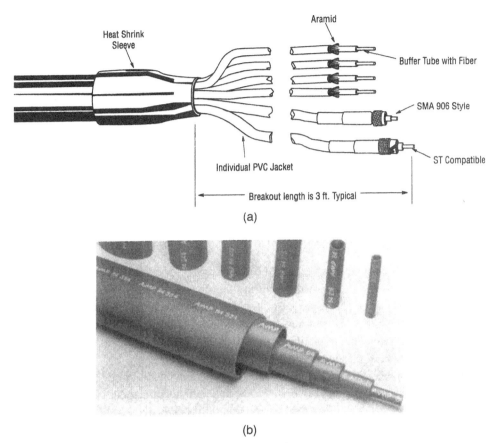

Heat Shrink
Sleeve

Aramid

Buffer Tube with Fiber

SMA 906 Style

ST Compatible

Individual PVC Jacket

Breakout length is 3 ft. Typical

(a)

(b)

Figure 4-21 Breakout of fibers from a multistrand cable.

ment locations. The jacketing is secured by heat shrink tubing. The tubing, which comes in many sizes, shrinks about 50% when heated. Heat may be applied with a soldering iron by rubbing the heated tip over the tubing, or with heat gun designed for the purpose.

The fiber-optic pigtail usually is thought of as a short length of cable, but it can be of almost any length. A pigtail typically is comprised of one or two fiber cables that have connectors on one end. The other end remains unterminated so as to be cut to length and spliced or terminated. Pigtails are most effective for equipment connection. Pigtail splicing has some advantage over unterminated cable with respect to labor saving of cable terminations. However, the cost of hardware protection may outweigh that advantage.

4.10 PREPARING A SPLICE AND TERMINATION

There are numerous types of connectors on the market with slightly different installation procedures. However, there are two installation steps that are common for all connectors.

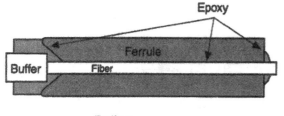

Figure 4-22 Epoxy application to bare fiber. (*Source:* Courtesy of Corning Cable Systems)

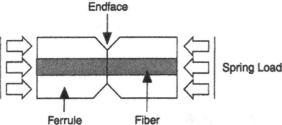

Figure 4-23 Physical contact splicing. (*Source:* Courtesy of Corning Cable Systems)

First, the fiber must be expoxied into the connector for long-term reliability of the connection. The epoxy allows polishing with fracturing the fiber and allows the fiber to be polished aggressively on the endface. The epoxy also seals the fiber from the environment and keeps to a minimum movement due to temperature changes. Figure 4-22 shows that the epoxy must coat the entire length of the fiber, around the buffer where the fiber enters the connector, and a bead must surround the fiber at the endface.

Second, the connector end surface must be polished and the fibers must have physical contact inside the connector adapter (TIA/EIA-568A) and hold under compression. Without physical contact between the ends of the fiber, attenuation of the light signal will result. Figure 4-23 depicts the proper mechanical physical contact installation of fiber in a connector.

There are several recommended fiber-polishing methods, depending on the fiber (ferrule) material used. Hard material generally is polished to be rounded on the endface and is referred to as preradiused. Softer material, such as glass-in-ceramic and thermoplastic, may be polished to a flat surface. The TIA/EIA-568A specifies a reflectance of < -20 dB for 62.5 μm and < -26 dB for single mode.

4.10.1 Fusion Splicing

Fusion splicing consists of aligning two clean (stripped or coated) fibers end to end and joining them by fusion with an electric arc. The ends of the fibers must be perfectly aligned by light injection and detection, a mechanical V-groove, or profile alignment. The alignment can be mechanical or automatic and assured with the aid of an optical power meter, viewing scope, or video camera. A simplified method is to use an optical power source and power meter. Figure 4-24 depicts the operation of a fusion splicer. As the ends of the fiber are brought together, an electrical arc melts and fuses the fiber. The splice must then be tested for attenuation loss.

Fiber fusion machines come in a wide range of features and prices. The high-end machines align the fibers, determine cleave quality, fuse the fibers, and test the connection.

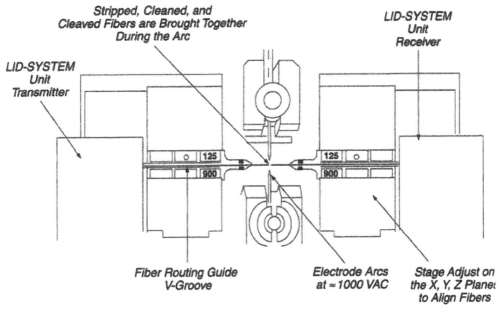

Figure 4-24 Fusion splicing with a fusion splicer. (*Source:* Courtesy of Corning Cable Systems)

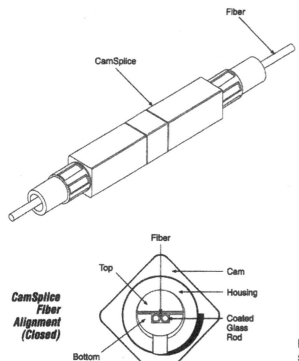

Figure 4-25 Nonadhesive mechanical splice. (*Source:* Courtesy of Corning Cable Systems)

4.10.2 Mechanical Splicing

Mechanical splicing utilizes a nonadhesive assembly approximately 2 in long to align the outer diameters of the fibers, assure the accuracy of core/cladding alignment, and achieve low splice losses. The fiber ends are cleaved, inserted into an alignment tube, and butted together. The tube contains an index-matching gel to reduce the attenuation losses at the fiber joint. The fibers are held together by compression, friction, or epoxy. Figure 4-25 depicts a self-contained mechanical splice device. The figure shows the alignment of the fibers in the container.

4.10.3 Mass Splicing

Mass splicing refers to the technique of splicing multiple fibers at once. Mass splicing can be accomplished mechanically or by fusion. Mass splicing is typically four to five times faster than single-fiber splicing. Fibers for mass splicing are typically formed into a ribbon cable configuration; however, that form of cable is not necessary. A device called a *fiber organizer tape applicator* (FOTA) can be used to prepare single loose fibers into a vertical flat ribbon cable. Proper preparation, cleavage, and fiber alignment are more important with multiple-fiber splicing than with single-fiber splicing. Although the TIA/EIA-568A standards allow a loss of 0.3 dB per splice for multimode fiber, today's technology allows for average losses of less than 0.10 dB.

4.11 PREPARING FIBER FOR SPLICING

Most manufacturers of connectors and splice hardware will furnish a step-by-step procedure for installation. Figure 4-26 illustrates a typical step-by-step procedure. The basic steps shown in Figure 4-26 are as follows:

  a. Cut the cable.
  b. Remove the jacket, buffer tubing, and other outer coatings with cutting pliers, using wire strippers that will not nick the fiber.
  c. Remove the plastic buffer coating mechanically or chemically (special chemicals are available). Mechanically the coating is removed with a precision type of wire stripper. The stripper must be set carefully so as not to damage the cladding. The safest stripper to use to prevent nicks in the fiber is a thermal type, shown in Figure 4-27. The stripper has a built-in battery to allow field operation. This item is also applicable for stripping copper wire and coax.
  d. Insert the sleeves in the cable.
  e. Glue the connector into place.
  f. Heat shrink the flexible sleeve.
  g. Crimp the metallic sleeve
  h. Place the connector into the appropriate holder (polishing tool) and polish the end of the fiber cable.
  i. Test the cable for light transmission and, if possible, for attenuation.

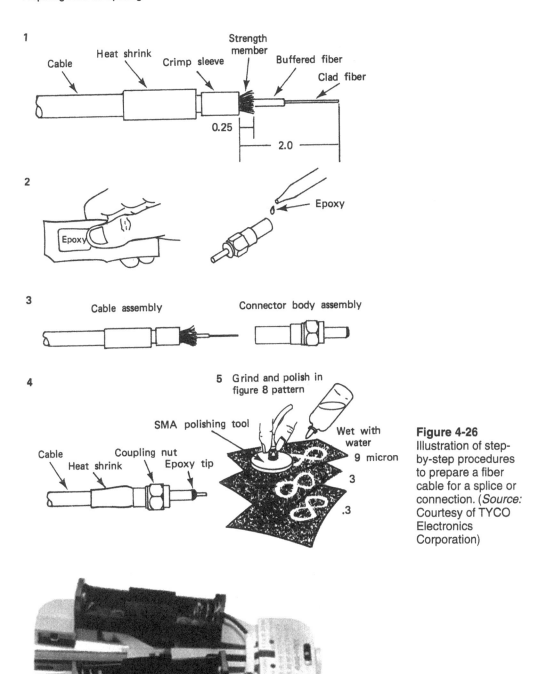

Figure 4-26 Illustration of step-by-step procedures to prepare a fiber cable for a splice or connection. (*Source:* Courtesy of TYCO Electronics Corporation)

Figure 4-27 An example of a thermal stripper.

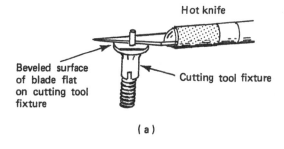

Hot knife

Beveled surface
of blade flat
on cutting tool
fixture

Cutting tool fixture

(a)

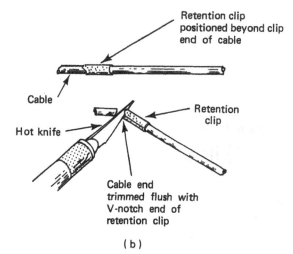

Retention clip
positioned beyond clip
end of cable

Cable

Hot knife

Retention
clip

Cable end
trimmed flush with
V-notch end of
retention clip

(b)

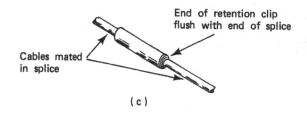

End of retention clip
flush with end of splice

Cables mated
in splice

(c)

Figure 4-28 Preparing a plastic fiber cable connection for termination in a connector.

Plastic cable is prepared in the same manner, except that the fiber cable must be cut with a sharp, hot knife to assure a flat end for polishing (Figure 4-28). Some success has been experienced by allowing a lens to form at the end of the fiber due to the flow of the acrylic as the hot knife cuts the material. The lens allows a good transfer of light energy without polishing the end of the fiber. Information concerning this technique can be obtained from the ESKA division of Mitsubishi Electronics America Inc. The author's students have tried the lens technique with varied results. The temperature of the cutting blade and the pressure of the blade on the plastic fiber seem to be critical in the development of a lens that is clear and will fit the end of the connector. The temperature of the blade determines the flow of the plastic, and the pressure deter-

1. Prepare cable ends to be spliced as shown in figure 1.

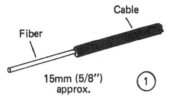

2. Insert fiber into cutting fixture/terminal block as shown in figure 2. Cut fiber using razor blade supplied, figure 3.

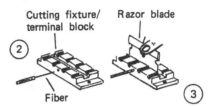

3. Dip the cut fiber ends into the silicone index matching gel supplied, figure 4.

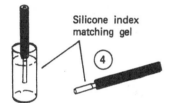

4. Slide protective shield onto the cable and insert fiber into the splice tube, figure 5.

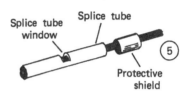

5. Crimp the splice tube. Fill the window with index matching gel and insert the other fiber into the splice tube, figure 6. Ensure that fiber ends are in contact.

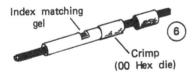

6. Crimp remaining splice tube end and slide the protective shield over the window. The splice is now ready for use, figure 7

Figure 4-29   Splicing a plastic fiber cable.

mines the length of time that the blade is in contact with the plastic. If the temperature is too high or the time that the blade is in contact with the plastic is too long, the lens will be oversized. This method seems to require a trial-and-error approach.

Figure 4-29 describes a simpler method of terminating plastic fiber. The method is recommended for field splicing of fiber cables and connectors for termination, and repair of breaks. The vendor, Nevada Western Corp., specifies a maximum of –1 dB loss per splice, and less than –3 dB loss for connection termination. Losses as low as 0.1 dB splicing loss and 0.3 dB connection loss have been reported by field technicians.

The installation or repair of fiber-optic cabling requires greater skill and more elaborate test equipment than required for copper media. The technician responsible for installing and

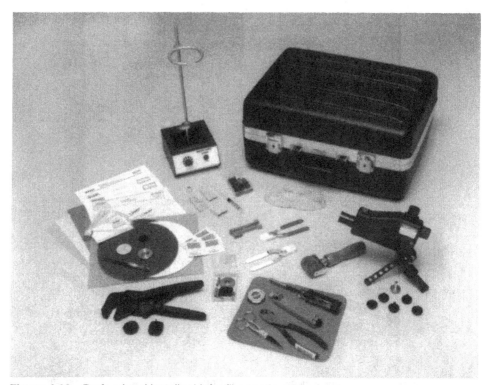

Figure 4-30   Professional installer kit for fiber-optic cabling.

maintaining fiber connections should possess an installer kit such as shown in Figure 4-30. The kit contains all the tools necessary to cut, polish, test, and splice epoxy-type fiber cabling.

Manufacturers are developing more efficient tools and easier methods of cutting and terminating fiber cable. The tool shown in Figure 4-31 is of special interest to the author. The set contains precision cutters to strip fiber-optic buffer and coating or to strip the insulation from copper stranded and braided wire. The unit utilizes heat from transformers for bench work or batteries in the handle for field work. The battery is said to allow for approximately 60 strips of fiber cable. Figure 4-31 illustrates the procedure for field stripping of either fiber or copper.

4.12 GENERAL CONSIDERATIONS

Although fiber splicing has been performed for several years, the technology is not static. Faster, less expensive, simpler, and more effective splicing equipment is entering the market. The chosen splicing method affects both splicing equipment and the hardware necessary to store and protect the spliced cables. The considerations when determining the most suitable splicing equipment are as follows:

Stripping procedures using Klein Micro-Strip™ Tools

The procedures described here are for stripping fiber-optic buffer and coatings with the MS-1-FS stripper. In the illustrations below, the fiber-optic jacket has already been stripped, (using the MS-2-L Handle Assembly, appropriate Cutter Blades and Tube Guides— see page 22) following the same steps described here for the buffer and coatings. Similarly, procedures for stripping 18-32 AWG wire and cables also are identical, using the MS-1 and MS-2 Handle Assemblies, the appropriate Cutter-Blade Sets and Tube Guides. Procedures for using Klein Soft-Strip thermal stripping tools are also similar, allowing 6-8 seconds in Step 2 for the heating element to soften the buffer/coating material.

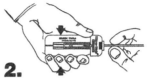

2. Close the handles all the way. Design of the tool components assures precision alignment of blades and fiber, for perfect concentric scoring without damaging the cladding or core.

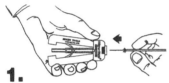

1. With the proper Cutter-Blade Set and Tube Guide seated in the MS-1-FS stripper, insert the fiber optic through the Tube Guide and into the Fiber-Support Channel through its orifice.

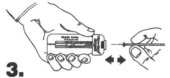

3. While maintaining hand pressure to keep the cutter blades closed, withdraw the perfectly stripped, undamaged fiber from the tool with a smooth, firm pulling motion. When the handles are opened, the stripped buffer/coating material will fall out. Should buffer or coating residue become lodged in the Fiber-Support Channel, simply remove the Fiber-Support Channel from the tool, remove the residue, and then reinsert the Fiber-Support Channel.

Figure 4-31   Stripping procedure. (*Source:* Kline Tools)

- Capital or rental outlay for equipment versus consumable expense
- Attenuation loss limitations
- Volume of splices annually
- Number of splicing crews deployed simultaneously
- Labor cost (contract or in house)
- Customer preferences
- Training for staff to become proficient
- Splice reliability
- Reflection limitations
- How the fiber will be terminated

4.13 FIBER-OPTIC PREMISE CONNECTIONS

The connecting hardware, such as racks, patch panels, and cross-connect panels, for fiber media is very similar to that used for twisted pair and coax. The difference, in an established campus, is that the fiber hardware is probably much newer and better organized than the copper hardware that was installed earlier. Figure 4-32 illustrates a main cross-connect closet for fiber-optic premise cabling with magnified views of the fiber, the cabling, the cabling connectors, and cross-connecting hardware.

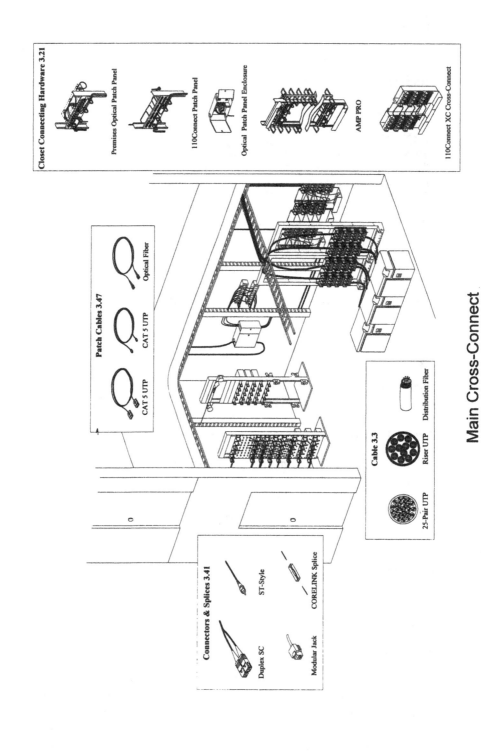

Closet Connecting Hardware 3.21

Premises Optical Patch Panel

110Connect Patch Panel

Optical Patch Panel Enclosure

AMP PRO

110Connect XC Cross-Connect

Patch Cables 3.47

CAT 5 UTP

CAT 5 UTP

Optical Fiber

Cable 3.3

25-Pair UTP

Riser UTP

Distribution Fiber

Connectors & Splices 3.41

Duplex SC

ST-Style

CORELINK Splice

Modular Jack

Main Cross-Connect

Figure 4-32 A main cross-connect closet. (*Source:* Courtesy of TYCO Electronics Corp.)

4.13.1 Campus Backbone

Fiber-optic cabling is used extensively and sometimes exclusively in campus backbones. A cabling network is established from a single hub or multiple hubs (cross connects) in the facility to the main cross connect in each building. Fiber can carry a complete spectrum of services: voice, video, and data. The outside plant cabling may contain multimode, single mode, or both.

4.13.2 Building Backbone

Building backbone connects the main cross connect in the building to one or more telecommunication closets. Optical fiber is becoming the choice for intrabuilding data backbone because of its ability to support multiple, high-speed networks in a much smaller cable without fear of cross talk or electromagnetic transmissions. New or remodeled facilities are using fiber exclusively. With the advent of xDSL the marriage of copper and fiber can be complete.

4.13.3 Horizontal Cabling

Fiber horizontal systems may be implemented for current requirements, such as FDDI[*], ATM[#], and Fibre Channel, and will easily accommodate systems such as token ring and Ethernet.

4.14 ADVANTAGES OF FIBER-OPTIC CABLING

Fiber-optic cabling should be considered for a new LAN system or addition to an existing system for the following reasons:

- **Structured cabling.** Fiber is the most attractive medium for structured cabling because it supports the widest range of applications at the highest speeds and greatest bandwidth, now and in the future. Fiber allows current systems to expand seamlessly in the future. Since fiber is immune to outside electromagnetic signals or noise and produces no electromagnetic emissions, it can share cables, conduit, and space with copper wires and cables.
- **Bandwidth.** The bandwidth of fiber-optic cable is over 1.2 Gbps per second, considerably higher than either coaxial cable or twisted pair. Optical cable transmits information by light energy instead of electrical energy and is therefore not subject to electrical properties of wire, such as resistance, inductance, and capacitance—properties that attenuate the signal and decrease the bandpass of a system.
- **Signal loss.** Signal loss in optical fiber is much less than in either coaxial cable or twisted pair. Signal loss for fiber cable is less than 8 dB per kilometer. This is approximately one-tenth that of coaxial cable. To compensate for attenuation loss an amplifier must be placed at 11 km as compared to the 1.1 km standard for coaxial cable.

- Fiber Distributed Interface is a 100 megabits per second fiber-optic network.
Asynchronous Transfer Mode is a very high speed technology.

- **Electromagnetic immunity.** The signal in fiber-optic cable is immune to electromagnetic interference (EMI) noise signals and radiofrequency interference (RFI). Fiber is any electromagnetic noise, such as cross talk, echoing, and ringing or static. Fiber cable operates noise free in factories, computer rooms, and other locations where electromechanical devices produce electromagnetic noise. Fiber produces no electronic emission and therefore is not a concern of the FCC or the European emissions standards.
- **Size.** Fiber-optic cable is considerably smaller than either coaxial or twisted-pair cable. In many cases fiber cables can run under a carpet without creating a ridge in a carpet and without the need for conduit as with copper cables.
- **Weight.** Fiber-optic cable weighs much less than either coaxial or twisted-pair cables. This can result in cost savings because the special hangers necessary for copper cabling are not needed.
- **Grounding.** Because fiber cables can be composed of all-dielectric materials, grounding concerns can be eliminated and lightning effects reduced dramatically.
- **Safety.** Fiber-optic cables carry no electrical energy so there is no possibility of an electrical spark. This allows fiber cable to be used in explosive environments, such as chemical plants and refineries. Since fiber cable has no metal, it cannot cause an electrical short circuit.
- **Security.** Fiber-optic cables offer far more security than twisted-pair or coaxial cables. There is no electromagnetic field, and it is almost impossible to tap into fiber cable without detection.
- **Adaptability.** Fiber cable is adaptable to LAN configuration, and with the proper connecting devices, fiber cable can be mixed with twisted-pair and coaxial cabling.

Table 4-10 summarizes the types of optical cable and their characteristics.

Each fiber in a multifiber optical cable is numbered and color coded. Table 4-11 gives the color codes for new trunk cable. Note that old cable may not have the same color code as new cable. Check the cable log. The outer jacket of optical cables identifies the type and function of the cable. Table 4-12 depicts the color code for the outside jackets of different cables.

4.15 SUMMARY

Most network wiring systems will have been developed over time and be a mixture of twisted-pair, coax, and fiber wiring. This requires the telecommunication technician to be schooled in the installation, testing, and repair of all three types. Figure 4-33 depicts such an example.

The wiring professional should balance the advantages of the three cable types against the cost of installation, cost of maintenance, availability, and future needs before selecting a cable type. It is possible that more than one type of cable will be the most cost effective for a particular campus installation. It is also possible that technologies such as infrared transmission, laser transmission, or microwave transmission might be used in some applications in place of cabling. All options (especially new technologies) should be examined before making a final plan. At

Table 4-10 Table of optical cable characteristics

Fiber Size (μm)	Description	Atten. Max. 850 nm (dB/km)	Atten. Max. 1300 nm (dB/km)	Bandwidth Min. 850 nm (MHz-km)	Bandwidth Min. 1300 nm (MHz-km)	Dim. Nom. (mm)	Cable Weight (kg/km)	Operating Temp. (°C)	Tensile Load Install. (N)	Bond Radius Min. @ Install. (cm)	Crush Resistance (N/cm)	Flame Rating[1]	Part Number
50/125	Light Duty Single, PVC	4.0	2.5	400	400	3.0	9.0	−20/+80	420	4.0	550	OFNR	502082-1[2]
	Heavy Duty Single	4.0	2.5	400	400	3.7	13.0	−20/+70	560	8.0	550	OFNR	501740-1[2]
	Light Duty Dual, PVC	4.0	2.5	400	400	3.0x6.1	18.0	−20/+80	840	4.0	550	OFNR	502085-1[2]
	Heavy Duty Dual	4.0	2.5	400	400	3.8x7.8	26.5	−20/+70	1120	8.0	550	OFNR	501116-5[2]
	Plenum Grade Single	4.0	2.5	400	400	2.8	6.5	−20/+80	420	5.0	450	OFNP	501819-1
	Plenum Grade Dual	4.0	2.5	400	400	2.8x6	13.1	−20/+80	840	5.0	450	OFNP	501693-1
	Breakout-2 Fiber	3.5	2.5	500	500	8.5	55.0	−30/+70	1800	17.0	700	—	501438-2
	HD Breakout-2 Fiber	3.5	2.5	500	500	9.4	70.0	−30/+70	4300	19.0	700	—	501498-2
62.5/125	Light Duty Single, PVC	4.0	2.5	160	500	3.0	9.0	−20/+80	420	4.0	550	OFNR	502083-1[2]
	Heavy Duty Single	4.0	2.5	160	500	3.7	13.0	−20/+70	560	8.0	550	OFNR	501739-1[2]
	Light Duty Dual, PVC	4.0	2.5	160	500	3.0x6.1	18.0	−20/+80	840	4.0	550	OFNR	502086-1[2]
	Heavy Duty Dual	4.0	2.5	160	500	3.7x7.8	26.5	−20/+70	1120	5.0	550	OFNR	501738-1[2]
	Plenum Grade Single	4.0	2.5	160	500	2.8	6.5	−20/+80	420	5.0	450	OFNP	501820-1
	Plenum Grade Dual	4.0	2.5	160	500	2.8x6	13.1	−20/+80	840	5.0	450	OFNP	501754-1
	DUALAN	4.0	1.5	160	500	4.75	20.0	−20/+80	1000	10.0	700	OFNR	501749-1[2]
	DUALAN Plenum	4.0	1.5	160	500	4.75	20.0	−20/+80	1250	10.0	700	OFNP	502024-1

Table 4-10 Table of optical cable characteristics (Continued)

Fiber Size (μm)	Description	Atten. Max. 850 nm (dB/km)	Atten. Max. 1300 nm (dB/km)	Bandwidth Min. 850 nm (MHz-km)	Bandwidth Min. 1300 nm (MHz-km)	Dim. Nom. (mm)	Cable Weight (kg/km)	Operating Temp. (°C)	Tensile Load Install. (N)	Bond Radius Min. @ Install. (cm)	Crush Resistance (N/cm)	Flame Rating[1]	Part Number
	Breakout-2 Fiber	4.0	2.5	160	500	8.5	55.0	−30/+70	1800	17.0	700	—	501438-4
	HD Breakout-2 Fiber	4.0	2.5	160	500	9.4	70.0	−30/+70	4300	19.0	700	—	501498-4
	IBM Type 5	6.0	4.0	150	500	7.5	48.0	−20/+80	1000	15.0	*	—	501714-1
	Light Duty Single, PVC	5.0	4.0	100	200	3.0	9.0	−20/+80	420	4.0	550	OFNR	502084-1[2]
	Heavy Duty Single	5.0	4.0	100	200	3.7	13.0	−20/+70	560	8.0	550	OFNR	501741-1[2]
100/140	Light Duty Dual, PVC	5.0	4.0	100	200	3.0x6.1	18.0	−20/+80	840	4.5	550	OFNR	502087-1[2]
	Heavy Duty Dual	5.0	4.0	100	200	3.8x7.8	26.5	−20/+70	1120	8.0	550	OFNR	501118-5[2]
	Plenum Grade Single	5.0	4.0	100	200	2.8	6.5	−20/+80	420	5.0	450	OFNP	501821-1
	Plenum Grade Dual	5.0	4.0	100	200	2.8x6	13.1	−20/+80	840	5.0	450	OFNP	501755-1
	Breakout-2 Fiber	4.0	3.0	200	300	8.5	55.0	−30/+70	1800	17.0	700	—	501438-6
	HD Breakout-2 Fiber	4.0	3.0	200	300	9.4	70.0	−30/+70	4300	19.0	700	—	501498-6
Single-Mode	Light Duty Single, PVC	—	1.0	—	—	3.0	9.0	−20/+80	250	4.0	550	OFNR	501530-1[2]
	Breakout-2 Fiber	—	1.0	—	—	8.5	55.0	−30/+70	1050	17.0	700	—	501556-1
	DUALAN	—	1.0	—	—	4.8	20.0	−20/+80	1250	10.0	700	OFNR	502119-1[2]
	DUALAN Plenum	—	1.0	—	—	4.8	20.0	−20/+80	1250	10.0	700	OFNP	502120-1

Flame Ratings[1]—Flame ratings for cables used within buildings are specified by the National Electrical Code (NEC). Underwriters Laboratories (UL), through the listing process, defines procedures to assure that products meet specific requirements.

General Use (OFN)—These cables pass the UL 1581 Vertical Tray Flame Test. They may be used for general installation in building wiring. They may not be used in risers or plenums unless installed in suitable conduits.

Riser Cable (OFNR)[2]—Riser Cables must meet requirements of UL 1666. and may be used in vertical passages connecting one floor to another.

Plenum Cable (OFNP)—Plenum cable must pass the Steiner Tunnel Test, UL 910. The cable may be installed in air plenums without the use of conduit.

Note: In all cases, cables meeting the more stringent requirements may be used in place of a particular cable.

118

Table 4-11   Color code for fibers within an optical cable

Fiber No.	Color	Fiber No.	Color
1	Blue	19	Red/Black Stripe
2	Orange	20	Black/Yellow Stripe
3	Green	21	Yellow/Black Stripe
4	Brown	22	Violet/Black Stripe
5	Slate	23	Rose/Black Stripe
6	White	24	Aqua/Black Stripe
7	Red	25	Blue/Black Dash
8	Black	26	Orange/Black Dash
9	Yellow	27	Green/Black Dash
10	Violet	28	Brown/Black Dash
11	Rose	29	Slate/Black Dash
12	Aqua	30	White/Black Dash
13	Blue/Black Stripe	31	Red/Black Dash
14	Orange/Black Stripe	32	Black/Yellow Dash
15	Green/Black Stripe	33	Yellow/Black Dash
16	Brown/Black Stripe	34	Violet/Black Dash
17	Slate/Black Stripe	35	Rose/Black Dash
18	White/Black Stripe	36	Aqua/Black Dash

Table 4-12   Outside jacket color code

Cable Family	Jacket Color
Bitlite	
10 micron (single mode)	yellow
50 micron	Gray
62.5 micron	Orange or Netural
100 micron	White
LanLite	
10 micron (single mode)	yellow
50 micron	Gray
62.5 micron	Orange
100 micron	White
Composites	Black
Breakout	
10 micron (single mode)	yellow
50 micron	Gray
62.5 micron	Orange
100 micron	White
Composites	Black
Loosetube	
All loose tube is black except industrial Tray Optics is orange	

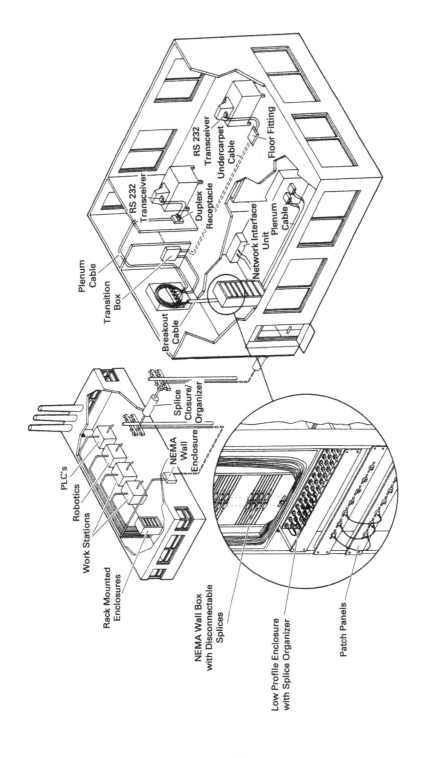

Plenum Cable

Transition Box

Breakout Cable

RS 232 Transceiver

RS 232 Transceiver

Duplex Receptacle

Undercarpet Cable

Floor Fitting

Network Interface Unit

Plenum Cable

PLC's

Robotics

Work Stations

Rack Mounted Enclosures

Splice Closure/ Organizer

NEMA Wall Enclosure

NEMA Wall Box with Disconnectable Splices

Low Profile Enclosure with Splice Organizer

Patch Panels

Figure 4-33 A factory LAN network.

present, twisted pair has gained in flexibility with the advent of DSL technology. However, in the future data will be transported by photons rather than by electrons.

QUESTIONS

1. What are the advantages of fiber-optic cable over twisted-pair and coaxial cables?
2. What are the disadvantages of fiber-optic cable as compared with twisted-pair and coaxial cable?
3. Why do fiber-optic cables offer more security of information than copper cables?
4. What type of fiber-optic cable is most appropriate when the attached equipment must be moved?
5. What is the purpose of a strength member in a fiber-optic cable?
6. Why is fiber-optic cable the best choice of cable in an electrically noisy environment?
7. What is the advantage of 50/125-μm fiber over 62.5/125-μm fiber?

NEC Requirements for Telecommunications Wiring

5.1 INTRODUCTION

The National Electric Code (NEC) is a set of guidelines issued by the NFPA (National Fire Protection Association) that minimize the hazards of electrical shock, explosions, and fires caused by electrical wiring. The guidelines establish the safe installation and operation of all wiring systems, including power systems, alarm systems, and communication wiring systems. These guidelines include the ampacity of wire, conduit capacity, grounding, and fire regulations. The NEC text is divided into nine chapters, and each chapter is comprised of individual articles. NEC codes cover private residences, hotels, apartment buildings, business buildings, factories, and all wiring and cabling that pass through any wall, floor, or ceiling and travel through conduits, ducts, plenum, or any air handling space (air conditioning).

Although the NEC is utilized by most wiring professionals, such as architects, the Organization of Electrical and Mechanical Engineers (OEM), wire product engineers, wiring contractors, and installers, each state, city, county, or other municipality has the authority to accept the NEC or establish its own codes. The NEC consensus standards are continually updated by the American Standards Institute. Proposals to amend or add to the NEC code may be submitted by any individual or organization. The specific format and deadlines for proposals may be obtained from the NFPA (address listed in the addendum). The deadline for proposals for the 2002 NEC standards was November 5, 1999.

5.2 NEC CABLE CERTIFICATION

Communication cables may be certified by Underwriters Laboratories, as UL/NEC, and by the Canadian Electrical Code C(UL)CEC as meeting the Bi-National Standards (BNS) CSA 22.2 No. 214/UL 444 and Section 60 of the Canadian Electrical Code, Part 1(CEC).

The C(UL) cable designations and meanings are as follows:

1. CMP—Cable meeting CSA FT 6 and UL 910
2. CMR—Cable meeting UL 1666
3. CMG—Cable meeting CSA FT 4
4. CM—Cable meeting UL 1581, Section 1160 (vertical trays)
5. CMX—Cable meeting UL 158, Section 1080 (VW-1)
6. CMH—Cable meeting CSA FT-1

The CSA flame test is defined in CSA C22.2 No. 0.3 listed in topic 8.0000.

5.3 NEC CATALOG REFERENCE INFORMATION

The NEC cable reference indicates the types of cable, their proper application, and flame resistance. Table 5-1 is a summary of NEC communication cable types, their application, and their properties.

Table 5-1   NEC cable reference information (*Source:* Courtesy of Belden Inc.)

New NEC Catalog Reference Information					
		Installation Type			
NEC Article/Type	**Description**	**Plenum**	**Riser**	**Commercial**	**Residential**
725 CL2	Class 2 cables	CL2P	CL2R	CL2	CL2X*
CL3	Class 3 cables	CL3P	CL3R	CL3	CL3X*
PLTC	A stand-alone class. This is a power limited tray cable—a CL3-type cable which can be used outdoors. Is sunlight and moisture resistant and must pass the vertical tray flame test.	(none)	(none)	PLTC	(none)
760 FPL	Power limited, fire protective signalling circuit cable	FPLP	FPLR	FPL	(none)
770 OFC	Fiber cable also containing metallic conductors	OFCP	OFCR	OFCG, OFC	(none)
OFN	Fiber cable only containing optical fibers	OFNP	OFNR	OFNG, OFN	(none)
800 CM	Communications	CMP	CMR	CMG, CM	CMX*
MP	Multi-Purpose Cables	MPP	MPR	MPG, MP	(none)
820 CATV	Community antenna television and radio distribution system	CATVP	CATVR	CATV	CATVX**

*Cable diameter must be less than 0.250".
**Cable diameter must be less than 0.375".

5.4 VERTICAL FLAME TEST

The following conditions apply to the flame test (FT-1), vertical flame test CSA C22.2 No. 03-92 Paragraph. 4.11.1. According to the guidelines, a finished cable shall not propagate a flame or continue to burn for more than 1 minute after five 15-second applications of the test flame. There shall be 15 seconds between flame applications. A flame test shall be performed in accordance with Paragraph 4.11.1 of CSA Std. 22.2 No. 0.3. If more than 25% of the indicator flag burns, the test cable fails.

The following conditions apply to the (FT-4) vertical flame test for cable in trays, CSA C22.2 Number 0.3-92 Paragraph 4.11.4. This test is similar to the UL-1581 vertical flame test, but is more severe. The FT-4 utilizes a burner mounted 20° from the horizontal with the burner ports facing upward. The UL-1581 test has its burner at 0° from the horizontal. The FT-4 sample must be larger than 13 mm (0.512") in diameter, or the cable samples must be grouped in units of at least three and bundled for an overall diameter of not less than 13 mm (2.219"). The UL-1581 vertical tray does not distinguish as to cable size. The FT-4 standard allows a maximum charring height of 1.5 m (59") measured from the lower edge of the burner face. The UL-1581 has a maximum allowable burn height of 1.98 m (78") from the burner.

The horizontal flame and smoke test per CSA C22.2 No. 0.3-92 Appendix B is for material that must pass a horizontal flame and smoke test in accordance with ANSI/NFPA Standard 262-1985 (UL-910). This standard states that the maximum flame spread shall be 1.50 m (4.92 ft). The smoke density shall be less than 0.5 at peak optical density and 0.15 at maximum average optical density.

5.5 LISTING, MARKING, AND APPLICATIONS OF COMMUNICATION CABLE

Most cable manufacturers designate a particular code for their cable that meets the flame test. The NEC states that communication cables installed within buildings shall be listed as suitable for their particular application and marked in accordance with Section 800-50 of the NEC. Communication cables are not to be marked with their voltage rating, since such marking may be misconstrued to mean that the cable is suitable for electrical power. Table 5-2 lists the letter designations for communication cable and the application of each from Section 800 of the NEC.

Cable designations (marking) indicate the application and flame and smoke limitations:

- **Type CM** cable is suitable for general-purpose communication use, with the exception of use in risers and plenums.
- **Type CMG** cable is suitable for general-purpose communication use, with the exception of riser or plenum applications, and shall be coded to be resistant to the spread of fire.
- **Type CMP** cable is suitable for use in spaces used for environmental air, such as ducts and plenums, and must be listed as having proper fire resistance and low-smoke emission.
- **Type CMR** riser cable is suitable for runs in a shaft or from floor to floor and must have fire resistance to prevent carrying fire from floor to floor.
- **Type CMUC** under-carpet cable and wire are suitable for under-carpet applications and must be listed as flame retardant.

Table 5-2   Cable marking and application

Cable designation	Classification	NEC reference
CM	General-purpose communication cable	51d and 53d
CMG	Communication general-purpose cable	1c and 553d
CMP	Communication plenum cable	51a and 53a
CMR	Communication riser cable	51b and 53b
CMUC	Under-carpet wire and cable	51f and 53d (exception No. 5)
CMX	Limited-use communication cable	51e and 53d (exceptions Nos. 1, 2, 3, & 4)
MP	Multiple-purpose general-purpose cable	51g and 53d
MPG	Multiple-purpose general-purpose cable	51g and 53d
MPP	Multiple plenum cable	Sec. 51g and 53a
MPR	Multiple-purpose riser cable	Sec. 51g and 53b

Source: Courtesy of Belden Inc.

- **Type CMX** is a limited-use cable that is suitable for use in dwellings and raceways and must be listed as flame resistant.
- **Type MP** is a multiple-purpose cable that meets the requirements of Types CM, CMG, CMP, and CMR cable. MP marked cable satisfies the requirements of NEC Section 760-71(b) (copper conductor in a multiconductor cable shall not be less than AGW 26 and single conductors shall not be less than AGW 18) for multiconductor cables and Section 760-71(g) for coaxial cable (coaxial conductors may have center conductors of 30% copper and shall be listed as Type FPLP, FPLR, or FPL cable).
- **Communication wires** such as for distributing frames and jumper wires shall be listed as resistant to flame.
- **Hybrid power and communication cable** shall be permitted where the power cable is Type NM performing to NEC 336 (factory assembled cable of two or more individually insulated conductors covered by an outer sheath with a minimum 600-V rating).

Figure 5-1 summarizes cable types, cable applications, fire resistant levels, test requirements, and the NEC code article applied to each application.

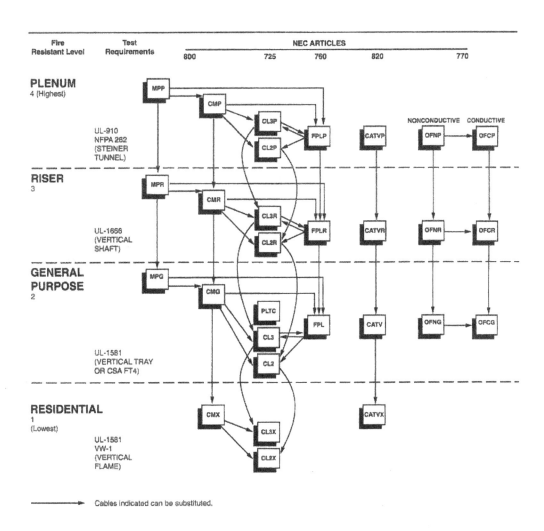

Fire Resistant Level	Test Requirements	NEC ARTICLES				
		800	725	760	820	770

PLENUM
4 (Highest)

UL-910
NFPA 262
(STEINER
TUNNEL)

NONCONDUCTIVE   CONDUCTIVE

RISER
3

UL-1666
(VERTICAL
SHAFT)

GENERAL
PURPOSE
2

UL-1581
(VERTICAL TRAY
OR CSA FT4)

RESIDENTIAL
1
(Lowest)

UL-1581
VW-1
(VERTICAL
FLAME)

⟶ Cables indicated can be substituted.

TYPES: MPP, MPR, MPG, MP = Multipurpose Cables

TYPES: CMP, CMR, CMG, CM, CMX = Communications Cables

TYPES: CL3P, CL3R, CL3, CL3X, CL2P, CL2R, CL2, CL2X = Class 2 and Class 3 Remote-Control, Signaling and Power Limited Cables

TYPES: FPLP, FPLR, FPL = Power Limited Fire Alarm Cables

TYPES: CATVP, CATVR, CATV, CATVX = Community Antenna Television and Radio Distribution Cables

TYPES: OFNP, OFNR, OFNG, OFN = Nonconductive Optical Fiber Cables

TYPES: OFCP, OFCR, OFCG, OFC = Conductive Optical Fiber Cables

TYPE: PLTC = Power Limited Tray Cables

National Electrical Code and NEC are registered trademarks of the National Fire Protection Association, Inc., Quincy, MA.

Figure 5-1 Summary of NEC cable applications and allowable cable substitutions. (*Source:* Courtesy of Belden Inc.)

5.6 INSTALLATION OF COMMUNICATION WIRES, CABLES, AND ACCESSORIES

The NEC Section 800-52 details the electrical and fire code requirements for the proper installation of communication wires, cables, and equipment. The highlights of the code are as follows:

• Communication wire or cables shall not be installed in any conduit, raceway, outlet box, compartment, receptacle, or junction box with electrical power or light circuits (exceptions are where the power circuits and communication circuits are completely separated by a barrier, or where the power circuits are solely for the purpose to supply power to the communication equipment, or for the connection of remote equipment, in which case the power and communication circuits shall be separated by a minimum of 6.35 mm, or 0.25 in.

• Communication wiring installations in hollow spaces, vertical shafts, or air ducts shall be made so that the spread of fire or smoke shall not be increased extensively. Fire stops shall be employed where communication cables penetrate fire-resistant walls, partitions, floors, or ceilings.

• Communication wires or cables shall not be attached to wiring conduits or raceways for support. Proper hangers or strapping must be attached to separate supports.

5.7 APPLICATIONS OF COMMUNICATION WIRES AND CABLES

The NEC codes specify the particular type of cable to be installed in areas that are the most hazardous for fire- or smoke-related conditions. A summary of the code follows:

• Cables installed in environmental air spaces such as ducts and plenums shall be Type CMP. The exceptions are Type CMG, CMP, CMR, and CMX wire or cable installed as per Article 300-22 and Section 645 of the code. Article 645 covers power supply wiring, equipment wiring, equipment interconnecting wiring, and data equipment rooms.

The communication room must meet stringent standards and include controlled access, main disconnect for equipment and heating and air conditioning (HVAC) power, fire extinguisher, fire-resistant walls, and fire-resistant door (Figure 5-2).

5.8 SUMMARY

Communication cable must be capable of withstanding the rigors of the particular environment in which it is installed. Furthermore, to prevent the spread of fire, such cable must meet the NEC fire code prescribed for its application.

Safety is the responsibility of every communication professional. The network designer must select the correct cable for the environment. However, the installer is ultimately responsible for installing the correct cable type.

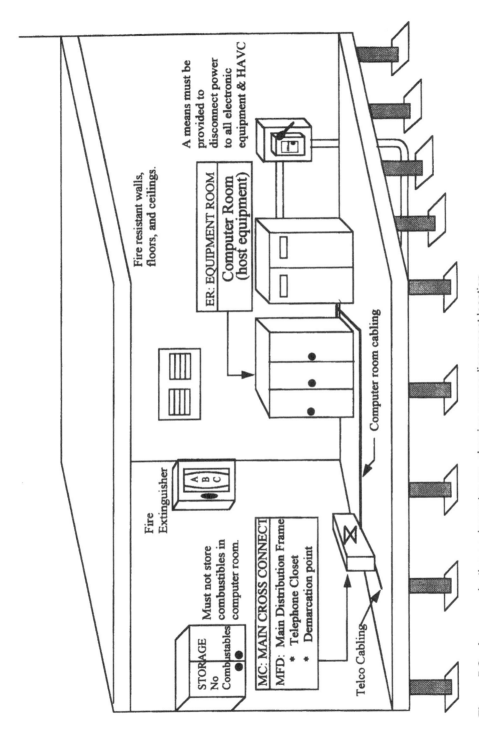

Figure 5-2 A communication equipment room, showing power disconnect location.

QUESTIONS

1. What is the NEC and how are codes modified or added?

2. Can local fire codes differ from the NEC?

3. What is the NFPA?

4. Who is responsible to see that cables are installed according to the NEC fire codes?

5. What are the organizations that test for code certification of cables and wires?

CHAPTER 6

Basic Network Topologies

6.1 INTRODUCTION

Network topology is the basic design or configuration of an information network that works together as a unit. Network topology describes how the various devices of the information network are connected to each other. The particular needs of the organization determine the type or types of network that can provide the most efficiency and cost effectiveness. This may prove to be a single network or a combination of different types of networks. The connections between devices are called *communication links* and can be phone lines, private lines, fiber-optic cables, coaxial cables, satellite channels, and so on.

A *local area network* (LAN) is a system, confined to a specific area, that allows all the computer hardware and software to work together. The LAN allows the use of PCs, Apples, mainframes, mini-computers, printers, and other peripherals to operate as a unit. The LAN combines resources, which should reduce cost and enhance efficiency. The computer equipment in most industries has grown in a haphazard way, with each workgroup or individual purchasing on a need or want basis and justifying the purchase by a show of immediate need. Little or no thought was given to its relationship with the resources of other departments or the company as a whole. At some point upper management looks at the overall cost of this unplanned growth and decides that it is time to consolidate. The problem then becomes one of intertwining all these mini-empires into a single functioning unit.

6.2 TOOLS OF THE TRADE

Before discussing the different types of networks the jargon of the network products must be understood. A few of the important terms are presented here. (A detailed glossary of telecommunication terms is presented at the end of the text; the list was as current as possible at the time of publication, but communication engineers continually add new words to the language. Refer to the glossary for definitions of unfamiliar terms.)

Network workstations are the terminals or computers that users use to access the network resources. Network workstations are DOS, OS/2, Macintosh, Apple, or network specialized workstations.

Network servers are computers that run the network operation system and manage the central stored data. These devices require automatic back-up, security (access and data loss), and large amounts of storage and memory.

Network interfaces or network adapters are plug-in cards that allow the PC to interface to the network and "speak its language." Two popular network adapters are Ethernet and token ring. Network adapters may have additional memory that allows for faster interface with the network by making internal decisions.

Network operating systems are the software that allows two or more workstations to access shared resources in a transparent manner. The operating system ties the workstations together as if they were peripherals to a mini-computer and affords services such as reliability, security, centralized management, and expanded resources.

Gateways connect networks with different protocols by changing the protocol of one network to the other. This is accomplished without any change of the data transmitted between the networks and at a speed that does not affect system operation. Gateways allow management to monitor system traffic.

Repeaters extend the length or area of a LAN by rejuvenating the data levels that are attenuated over the wiring media. Repeaters amplify all the signals carried by the network without making any modifications or inducing signal distortion or error.

Bridges are intelligent repeaters in that they not only connect networks and are transparent but also serve a number of additional functions. Bridges can direct traffic by selecting and directing data packets, provide security access with password protection, and provide statistics for management control. Bridges can also be used to direct traffic over different wiring media. Bridges are relatively inexpensive and easy to install.

Routers serve the same functions as bridges of connecting LANs or WANs (wide area networks) and, like bridges, are intelligent repeaters. They also analyze cost, time, and priority to determine the most efficient path for data traffic. Routers are more difficult to install and more costly than bridges. For this reason, routers should be used where they are the most effective—for interconnecting remote locations (WANs) with large amounts of traffic that may cause congestion.

Brouters offer the range of options of network management—wide protocol support, the ability to determine the most efficient path for packet travel over the system—of both routers and bridges. Their greater versatility, of course, results in a high cost and more difficult installation.

Processing power is the speed and memory capability of the computer (or the system in case of a LAN).

System memory is the total memory capability of the system. The memory is usually thought of as disk drive storage or the total storage capability of all the disk drives on the network.

Data storage is the storage of data on a disk drive or on a disk drive and in ROM (read-only memory).

Back-up is the method by which a system protects data and data transfer from loss by any means. Back-up can also mean the capability of a system to reroute in case of transfer media failure. It is impossible to anticipate every possible failure occurrence. However, a company whose life depends on telecommunication can be out of business without back-up that can recover from catastrophic failures (for example, the companies that depended solely on the public utilities during the 1989 California earthquake).

Modems are devices that convert digital signals to analog signals to be transmitted over telephone lines. Modems may be built internal to a printed circuit board or in an external unit. Modems were once required to transmit digital information over telephone lines. However, telephone companies are rushing to install digitally compatible lines on which data processing information can be transported directly and telephone audio signals are converted to digital signals for transmission. Modems are available in baud rates (transmission rates). Currently, modems with speeds of up to 56,000 baud are available.

Codecs are devices that convert audio signals to digital signals to be transmitted on a telephone line. The advantages are less noise and better sound quality.

Direct service line (DSL) is the latest technique to transmit high-speed audio, video, and data over twisted-pair wiring.

The author has included a Glossary of several thousand words pertaining to the communication industry at the end of the text.

6.3 BASIC NETWORK MODELS

The two basic types of network models that describe most networks in use today are **point-to-point** and **multipoint systems** (Figure 6-1).

Figure 6-2 depicts a simplified LAN network that might be used by an office network or a department in a college.

6.4 DETERMINING NETWORK CONNECTIONS

Network topology is determined by interconnection needs. For example, a group of devices that need to talk to each other are referred to as *stations*. The stations may be computers, terminals, modems, printers, or other devices. Each station attaches to a network *node*. We must determine the type of topology that could best interconnect the stations. One solution would be to connect each station to every other station using point-to-point links. Obviously, there are some major

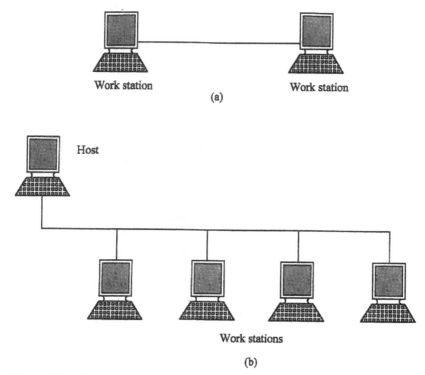

Figure 6-1   Two basic network models: (a) point to point, (b) multipoint.

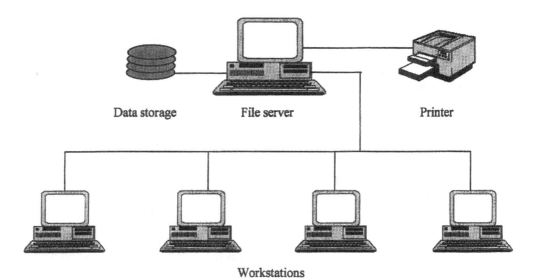

Figure 6-2   Simplified LAN.

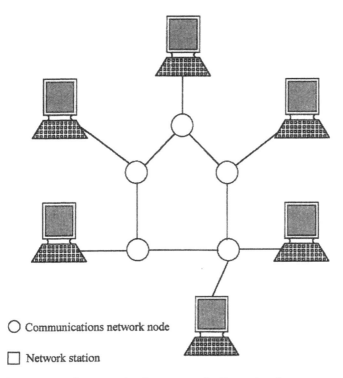

○ Communications network node

☐ Network station

Figure 6-3   An example of a communication network.

problems with this type of approach and perhaps some impossibilities as well. Another, better solution to the problem of interconnection is the use of a *communication network* that is capable of transferring data between stations (Figure 6-3).

A communication network provides transmission facility between stations. Several types of networks have been developed to meet different needs, and all have advantages and disadvantages. We will consider these networks and their advantages and disadvantages in the following sections.

6.5 POINT-TO-POINT TOPOLOGY

At the basic level, data communication takes place between two devices that are connected directly. A simple example of this is two cans connected via a string. Once the string is taut, a child at one end holds a can to his or her ear, thus hearing what is being said at the other end. For many this may have been their first experience of *point-to-point transmission*. An example of point-to-point communication in computer systems was shown in Figure 6-1. Point-to-point nodes only communicate to other directly connected nodes. Sometimes the nodes are physically located next to each other, but this is not necessary. The two nodes could be located a distance from each other, perhaps in different rooms.

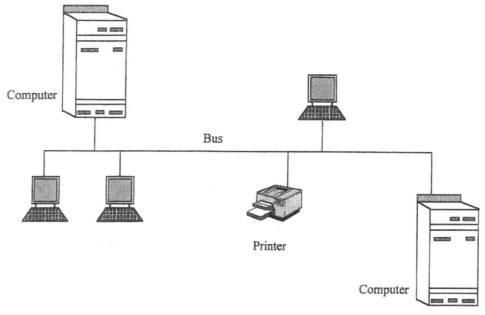

Figure 6-4   Bus network.

6.6 *MULTIPOINT OR MULTIDROP NETWORKS*

A multipoint network (Figure 6-1b) is one in which more than one node shares one line by shar-ing time on the line. This is called *time sharing*. Originally, this type of network was designed for high-speed data transmission rates or where there was a large volume of communication traf-fic over the network. However, there are now several types of multipoint networks, including the bus network. Many companies use these systems to automate their production facilities because they can be installed quickly and are very cost effective. Typical networks are the *bus network* (Figure 6-4), the *star network* (Figure 6-5), and the *hierarchical network* (Figure 6-6). Each of these networks will be explained in the following sections.

6.7 *BUS NETWORK*

A bus network (Figure 6-4) has a common communications medium to which multiple nodes are attached. The bus network is a special case of a tree topology (Figure 6-7) that is characterized by having only one trunk and no branches. The tree can have multiple modes (in this case PCs, workstations, a printer, and so forth). Each component must have its own interface device. This interface device is usually a card, software, and hardware to access the network. Since multiple devices share a single data path, there must be a mechanism to determine which node will obtain the right to transmit first. This mechanism is called the *access method* and will be discussed later in this chapter.

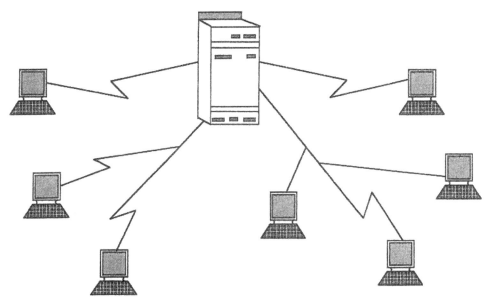

Figure 6-5 Star network.

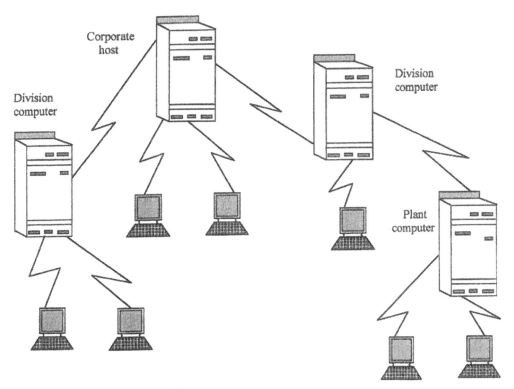

Figure 6-6 Hierarchical network.

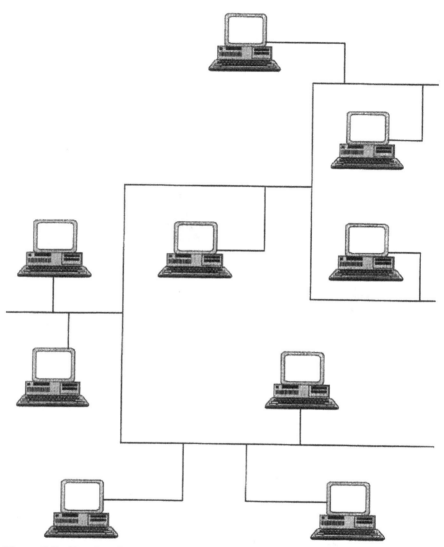

Figure 6-7 Tree topology.

Since the nodes share a single data path or bus, each device must have a unique address. The data are sent along the bus until the address of a specific device is found. Bus implies high speed, and bus networks are usually implemented in situations where the distance between all nodes is limited (e.g., a building or a department located in the same building). The loss of a single node on a bus has little or no impact on the other nodes unless the entire bus fails. Therefore, the reliability of this type of network is excellent. Each device is connected to the bus by a *tap connection* that breaks into the bus cable. However, tap connections cause a certain amount of signal power loss on the cable. Therefore, only a limited number of devices can be connected to them.

Another disadvantage with bus-type connections is problem isolation. Since all devices are serially connected to the bus, each device must usually be tested in sequence to locate a problem. A typical implementation of this type of network is on LANs.

6.8 STAR TOPOLOGY

In the star network that was shown in Figure 6-5 a central switching network (typical of a host computer) is used to connect all the nodes in the network. The circuit can be point to point, multipoint, or a combination of one or more of these topologies. Such network configurations are called *hybrid networks*. Examples of these networks are shown in Figure 6-8. This arrangement makes it easier to manage and control the network than with some other configurations. The disadvantage of this type of network is that since all data flow through the central node, a failure causes the entire network to go down. Another potential problem with this design is that at peak periods the central node may become overloaded and unable to transmit.

There is no limit to how many arms can be added to the star network, nor is there a maximum length of these arms. Therefore, it is easy to expand a star network by adding more nodes. When a central computer supports many terminals, the star configuration is easy to implement. Since the central computer is the master node and controls the network, the rules for the network operation are relatively simple. A typical application of the star network is the dial-up telephone system, in which the individual telephones are nodes and the public branch exchange (PBX) acts as the central controller.

6.9 HIERARCHICAL TOPOLOGY

The hierarchical network (Figure 6-6) is sometimes known as a tree structure (Figure 6-7). The top node of the structure is called the root node. This type of network would most likely be implemented where the lower-level nodes at the second or third level are in themselves computers. One advantage of this type of network is that even if the root node fails, the network would stay up because there are computers at the lower hierarchy. Many of the features in the bus topology are shared with the hierarchical network. Many networks contain a combination of several other networks.

6.10 RING NETWORK

Ring networks (Figure 6-9) consist of a closed loop. Each station is connected to two other stations, forming a ring or a circle. All communications follow a clockwise or counterclockwise rotation. Most stations in ring networks are close together, usually in the same department or room of a building. The information that is passed along the ring is in the form of a data packet. The data packet that is sent by the originating station (source node) contains source destination (destination mode) and data fields. As the packet is circulated around the ring, a receiver/driver

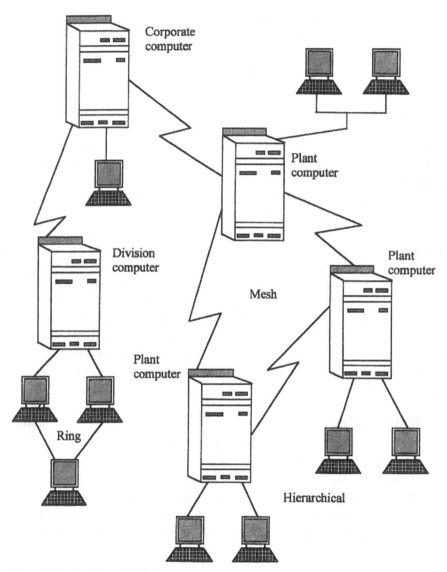

Figure 6-8 Hybrid networks.

(sometimes called a repeater) in each device checks the destination address of the incoming packet and either routes it to the station itself by simply copying the data into a local buffer or sends it to the next station. This regeneration is important to note because it eliminates the typical attenuation that occurs during signal propagation and in bus networks. A classic ring design is really a series of point-to-point connections. However, there are many ring designs.

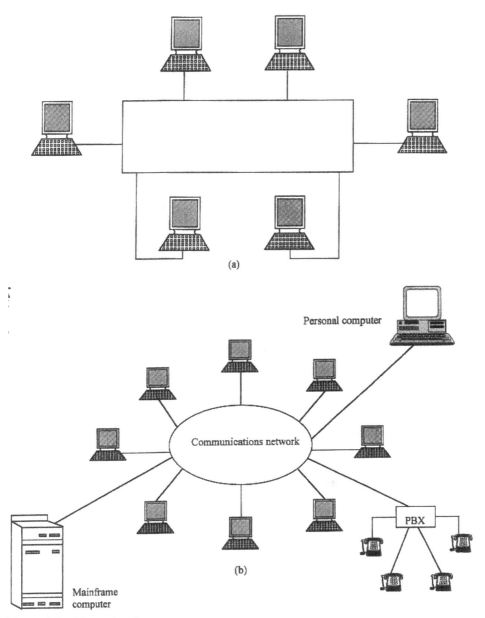

(a)

Personal computer

Communications network

PBX

Mainframe
computer

(b)

Figure 6-9   Ring network.

6.11 NETWORK ACCESS PROTOCOLS

We have discussed to a limited degree the architecture of several LANs. Each of these topologies has one common problem—that is, computer terminals, printers, application systems nd software manufactured by different vendors. However, great strides are being made to integrate connectivity between products.

To provide uniformity, network user organizations have established network protocols. The largest of these, the *International Standards Organization* (ISO), has established what is called the *Open System Interconnection* (OSI) standards (Table 6-1).

Table 6-1   Levels of OSI protocol

Application
Presentation
Session
Transport
Network
Data link
Physical

The functions of the OSI levels are as follows:

1. *Physical level.* A set of rules regarding hardware and transporting data across a medium. It monitors the transmission of electrical signals between two data communication stations.
2. *Data link level.* This level is involved in transmitting data frames that indicate the beginning and ending of a data frame. This layer also shields higher levels from the physical transmission media.
3. *Network level.* This layer routes data from one node to another. It takes data from the fourth level and assembles it into data packets, which are sent to the two lower levels for transmission.
4. *Transport level.* The transport level is responsible for recognizing and recovering from errors, selecting the class of service, monitoring transmission to measure service quality, multiplexing messages into one circuit, and then directing the message to the correct circuit.
5. *Session level.* The function of the session level is to control when users can send or receive data, verify passwords, and determine who transmits or receives and for how long.
6. *Presentation level.* The presentation level is concerned with presentation of information in a form that is meaningful to the user, network security, file transfers, and formatting functions. This layer may include code transmission, data conversion, data compression and expansion, and word processing. This layer establishes a table of transmission code that allows one terminal protocol to address terminals that utilize a different format.

7. *Application level.* The application layer performs functions that are user specific, such as establish and terminate connections between users and the main frame, and handling electronic mail, file-server and printer programs, and database management programs.

The OSI model was presented here as a sample of protocols. It is not within the scope of this text to discuss all telecommunication protocols.

6.12 MAKING THE RIGHT CONNECTION

Regardless of the size of your organization or the number of pieces of equipment involved, bringing your company's users together into a common local area network is no easy task. The network system must be transparent to all the user community so that users are able to operate as before but with the added advantage of the network's resources.

Some of the factors involved in developing an encompassing network are as follows:

• Connect the network or networks that your company has today into a system that will accommodate the expanding needs of the future.
• Choose a wiring system that meets the needs of today and can be expanded for future needs.
• Build a totally encompassing system that can be effectively managed and easily expanded for future needs.
• Survey the entire staff and management to determine present and future needs.

The key to a smoothly operating, completely user transparent LAN or WAN is *interoperability* of all the systems, hardware, wiring media, operating systems, and other software.

The wiring media for most large LANs will consist of a combination of twisted-pair, coaxial, and fiber-optic cabling. It is up to the wiring professional to make these different media function in a seamless marriage.

6.13 SUMMARY

Local area networks range from the very simple to very complex. To select the correct one for the corporation, the telecommunications professional must be well informed or seek the advice of an outside consultant.

QUESTIONS

1. Why is it important to establish standards for LANs?
2. What type of LAN is the Ethernet system?
3. What are the seven levels of the OSI protocol?

CHAPTER 7

Digital Subscriber Line

7.1 INTRODUCTION

Digital subscriber line (DSL) refers to modemlike technology that allows the transmission of voice, video, and data over existing copper telephone lines at megabit speed. DSL provides dedicated bandwidth that is over 278 times faster than a 28.8-kbps modem, 143 times faster than a 56-kbps modem, 62 times faster than ISDN,* and up to 4 times faster than a T1 connection. DSL uses existing telephone twisted pairs without tying them up. The lines can be used concurrently for voice, data transfer, and faxing.

DSL technology was originally developed to allow customers access to video over telephone lines. The market for video on demand did not develop, and the technology lay dormant until the Internet became accepted. The technology has now been rejuvenated and enhanced to the point where it may be the solution to many needs of the customer and problems faced by the provider. Low-cost wide bandwidth over copper pairs is a reality (for limited distances), and providers are scrambling to provide the service.

The field is evolving rapidly. The material presented here is the latest state of the art. DSL bandwidth, bit rate, and distance will probably increase with improvements in protocol, innovation, and equipment improvement. However, the basic concepts, definitions, and network solutions presented here will remain valid for some time.

The demand for faster, more intelligent, higher capacity, and cheaper networks is causing exceptional growth in nearly all sections of the communications market. There is no end to the insatiable corporate and public demand for faster and cheaper high-speed communications ser-

---

* The integrated service digital network (ISDN) comes in two basic types: BRI, which is 144 kbps and designated for desktop, and the PRI at 1.544 Mbps and designated for telephone switches.

vices. These facts increase the market and drive the entire communications community to develop protocols, hardware, and software that allow Internet service providers (ISPs) to open significant opportunities for innovative services. The new communications war will be between the ISPs with cable networks and ISPs who are xDSL providers over telephone company twisted-pair lines. The winner in this drama will be the consumer, both business and residential users. While xDSL offers wide bandwidth and other options, it has a decided limitation of distance. The 5-mile limitation to the central office (CO) eliminates all but densely populated areas. Sixty percent or more of residential users cannot, with current technology, be reached. However, DSL offers economical possibilities for business and private subscribers in metropolitan areas. Currently, the frenzy of effort of suppliers is toward DSL on copper cable, and to a lesser extent fiber. However, we should not completely discount wireless, especially for the rural residential market.

DSL service is expected to become extremely popular because it provides Internet access rivaling, and in many cases exceeding, the speed of T1 and other leased lines over twisted pair at a fraction of the cost.

Typically, businesses and residents had three options for access to the Internet: analog dial-up modems, ISDN connections, or leased lines. Analog dial-up modems connect the masses to the Internet. The monthly cost is low, but so is the speed. In a world where time is money, or so we think, speed is king. ISDN technology has proven to be a cumbersome technology for the telephone companies. Not only is it difficult to install, but also the 144-kbps end user rate is slow, considering the monthly cost and metering charges. Leased lines, such as T1, have become the most popular solution for businesses communicating over the Internet. However, the $1000 to $2000 monthly fees are often beyond the budget of most small businesses. DSL offers business and residential customers more diverse and faster networks at a fraction of the cost of leased lines. There are a number of flavors of DSL available. The term xDSL is used to cover the spectrum of DSLs available.

7.2 DSL DESIGNATIONS—xDSL

There are several transmission techniques available to transmit high-speed digital data over copper wires. These include two binary one quaternary (2B1Q) modulation, discrete multitone (DMT) modulation, and carrierless amplitude phase (CAP) modulation. Each of these has its pluses and minuses. However, they are all capable of high-speed digital transmission over copper. A number of "standards" have emerged from various entities. Some of these are designed to operate over a single copper pair, and others require a double copper pair.

DSL comes in several varieties that are umbrellaed under the term xDSL. The following paragraphs provide some insight into the several varieties of DSL.

ADSL (asymmetric digital subscriber line) is often referred to as ADSL—full rate, G.dmt, and G.922.1. This ADSL ANSI standard provides a voice connection and data connection on the same customer local loops that currently support only one analog voice connection. The data connection supports from 64 kbps to 1 Mbps upstream (to customer premise) and 1.5 Mbps to 8

Mbps downstream. ANSI is a U.S. standards organization that cooperates with governmental, industry, and vendor groups and associations to develop voluntary standards in many areas, including communications. ANSI also participates in the International Standards Organization (OSI) to develop voluntary international standards. The asymmetrical aspects of ADSL make it ideal for Internet surfing, video on demand, and remote LAN access as users of these applications download more information than they send. Due to its asymmetric nature and ability to coexist with analog phones, ADSL has been touted as the favored technology for residential applications. However, ADSL has run into problems regarding cost and distance limitations. ADSL can also cause interference to other services, such as T1, running on the same bundle. ADSL is less appealing to business users as they often need to upload as much data as they download from a network and the ability to share a single copper pair with analog telephone is much less valuable to them.

ADSL requires a voice/data or POTS (plain old telephone service) splitter to be installed at the customer's location to separate voice and data signals. Full rate ADSL is currently up to a maximum range of 18,000 ft (3.4 miles, 5486 m, or 5.5 km) from the provider's office to the enduser. ADSL Lite technology, called Universal, provides service ADSL, G.shdsl, Splitterless, or G922.2 and does not need a splitter to be installed at the customer's premise.

HDSL (high-bit-rate digital subscriber line) delivers high-speed data networking up to 1.554 Mbps over two copper pairs and up to 2.048 Mbps over three copper pairs to a range of 20,000 (3.8 miles or 6.1 m) from the CO. HDSL is similar to SDSL and has symmetrical transmission capabilities. Most T1 lines in use utilize this technology.

HDSL2 is an application-specific technology designed to transport T1 service (1.544 Mbps) data over a carrier service area (CSA). HDSL2 was approved by IEEE Standard T1E1.4 and published as ANSI T1418.2000. It is a two-wire replacement for the four-wire HDSL. HDSL2 provides a single-pair T1 transport with near flawless interoperability. This standard offers high-bit-rate symmetrical DSL over one copper pair and increases distance reach and data rates. The proposed standard for G.shdsl allows service providers to offer data rates from 192 kbps to 2.3 Mbps to customers up to 20,000 feet (6096 m) from the central office.

Table 7-1 compares HDSL to HDSL2.

G.shdsl (G.992.2 SHDAL) is the first standardized multirate symmetric DSL and is the product of the International Telecommunication Union (ITU-T). G.shdsl is designed to transmit symmetrical data across a single copper pair at data rates of 192 kbps to 2.304 kbps. This covers applications traditionally served by T1 through E1 services. The new standards proposed by G.shdsl will increase bandwidth and distance of service over a copper pair and overcome some of the distance problems with ADSL. Data-focused competitive local exchange carriers may be able to solve, to some degree, the distance problem with G.shdsl.

Four-wire operation of G.shdsl extends the bit rate to 384 kbps to 4.720 Mbps and the distance to 24,000 feet (7315 m). This is a significant improvement over the current DSL limitations of 1 to 1.5 Mbps over a distance of 10,000 feet (3048 m). Figure 7-1 compares the distance versus data rate of SDSL to the proposed G.shdsl. The slower speed of G.shdsl should not be a

Table 7-1   Comparison of HDSL to HDSL2

HDSL	HDSL2
Uses two pairs of copper wire, with each pair running 784 kbps upstream and downstream	Uses one pair of copper wire and delivers 1.55 to 2 Mbps
Tolerates mixed bridge taps and gauges	Tolerates mixed bridges and gauges
Cross talk; inability to serve similar services on the same cable pair	Mixed cross talk performance at less than 5 dB. Suitable for use in worst-case environments
12-kft coverage, a full carrier area	12-kft coverage, a full CSA
Latency more than 500 μs	Latency less than 5 μs
Cross talk interference	
Mostly proprietary, based on ANSI and Bellcore[*]	Vendor interoperability between HDSL2, termination units, line termination units, network termination units

* Bell Communications Research formed to provide centralized service to seven regional Bell Holding Companies.

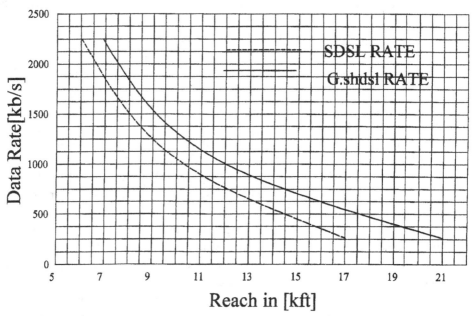

SDSL Data Rate vs. Reach

Figure 7-1   Comparison of distance versus data rates to SDSL and proposed G.shdsl.

problem with residential PC users who need a much faster download than upload. The main advantage of G.shdsl over other ADSL deployments, other than distance, is that it can be splitterless and not create interference with digital signals. The G.shdsl standards should allow the reduction of hardware such as repeaters because the signal is stronger and can travel a greater distance. The G.shdsl standards are currently being reviewed by the ANSI and the ITU-T. It is hoped that approval by these standards organizations will urge the adoption of G.shdsl as an international standard.

G.shdsl takes SDSL to a new level by utilizing a different line code, called trellis-coded postamplitude modulation (TC-PAM). TC-PAM is the modulation format that is used in both HDSL2 and G.shdsl and provides excellent performance over a variety of loop conditions. G.shdsl uses TC-PAM to provide far reach/high rate capacity, offering enhanced performances of increased reach and rate and improved bandwidth capacity over ADSL. Compared with HDSL2, G.shdsl offers lower power consumption and intelligently shaped transmit waveforms. The TC-PAM technique is used in 56-kbps modems and was adapted for HDSL2 to allow HDSL services over a pair of copper wires. Optional repeaters are defined for both single-pair and double-pair operation.

While G.shdsl approval is not expected until the end of the year 2000, equipment manufacturers of HDSL2 services should be able to make easy transition to G.shdsl standards.

SHDSL (symmetric high-bit-rate digital subscriber line) provides high-speed data transfer over a single pair of copper phone lines. SHDSL provides speeds of 16 kbps up to 1.544 Mbps both upstream and downstream for a range of 24,000 feet (4.5 miles, or 7.2 km). SDSL is ideal for business applications that require the same speed for both upstream and downstream transmission, such as collaborative computing and video conferencing. SDSL uses the type of line-modulation techniques, known as 2B1Q, employed in ISDN and has been in use for a number of years as high-bit-rate DSL (HDSL) to provide T1 service. SDSL is known to be highly compatible with other services running on the same bundle.

Table 7-2 summarizes the key facts about SHDSL. The primary advantage of SHDSL is that it is available now in many locations. Table 7-3 summarizes some of the applications for SHDSL. Table 7-4 compares the characteristics of HDSL, SDSL, HDSL2, and G.shdsl.

IDSL (IDSN digital subscriber line) provides symmetric upstream and upstream speeds of 64 to 144 kbps over a single copper pair. The range of IDSL from the CO is 18,000 ft (3.4 miles or 5.5 km). The distance can be doubled with a midspan U-loop repeater. IDSL uses the same kind of line-modulating technique as SDSL and ISDN.

VDSL (very high-bit-rate subscriber line) is currently the fastest xDSL technology, delivering downstream 13 to 52 Mbps and upstream 1.5 to 2.3 Mbps over a single copper pair with a range of 1000 to 4500 ft (0,1 0.9 miles or 1.67 to 2 km) from the CO. The range depends on the bit rate.

Table 7-2   Key facts about SHDSL

Two or four copper pairs

Variable bit rate from 192 kbps to 2.32 Mbps

2.4-km loop length at 2 Mbps and 4.5-km loop at 384 kbps

Granularity of 8 kbps two lines

Granularity of 16 kbps four lines

Flexible SHDSL framing

Synchronous transmission

Asymmetric option at certain data rates

Annex including HDSL2 features

Table 7-3   SDSL services

For business subscribers

Multiple data and telephone channels

Video conferencing

Remote LAN access

Web page hosting

Live audio transmission at Quality of Service (QoS)[*] rating

Leased lines with specific data rates

The same level of bandwidth upstream and downstream

For residential users

Fast Internet and telephone access over one line

Internet games

Remote home to business connection

* QoS is the standard of telephone audio.

Table 7-4   Comparison of symmetrical data technologies

Comparison of Symmetrical Data Technologies				
	HDSL	SDSL	HDSL2	G.shdsl
DATA RATES	T1	192kbps-2.3Mbps	T1	192kbps-2.360Mbps or 384kbs-4.720Mbps
Pairs	2	1	1	1 or 2
Line Code	2B1Q	2B1Q	TC PAM	
EOC	13-bit Triple Echo	Propnetary	24-bit HDLC-type	20-bit HDLC-type
Pre-activation	None	Propnetary	Standard Based	G. 994.1
Rate Adaption	No	Yes(?)	No	Yes
Power Back-off	No	No	Yes	Yes
Repeaters	Yes	No	Yes	Yes
Timing	Piesiochronous	Synchronous	Piesiochronous	Both
Span Power	Yes	Yes	Yes	Yes
Reach:				
2.304Mbps	na	Less than CSA	na	6kft
1.5444Mbps	CSA	Less than CSA	CSA	CSA
784kmbps	na	CSA	na	RRD (11.5kft)
384kbps	na	CSA+	na	RRD (15kft)

7.3 xDSL APPLICATIONS

During the first wave of DSL deployment, service providers are building regional and national networks and providing high-speed Internet and wholesale DSL access. Start-ups are often offering free connection to new customers.

Figure 7-2 denotes the evolution of subscriber lines and access. Figure 7-2a illustrates the line requirements for E1/T1, HDSL, and SDSL. Figure 7-2b compares the evolution of subscriber access speed. Note that SDSL is far faster upstream than other protocols.

7.3.1   Voice Over DSL (VoDSL)

The ability to offer wide bandwidth via DSL allows concurrent voice and data over one copper pair. This cost-saving innovation is driving rapid market growth. The industry is moving toward cost-saving applications at the desktop that combine Internet access services and multiple voice lines. Internet service providers are offering intelligent Internet provider (IP)-based services with guaranteed quality of service (QoS) at premium prices to develop a customer's base.

Figure 7-3 illustrates an example of a home or small business network connected to an xDSL line, which can be a single copper pair, a double copper pair, a coaxial cable, a fiber-optic cable, or wireless. The home PCs may be interconnected by any of these media or by high-frequency signals over the power line. Any of the PCs can use any of the peripherals, and two or more users may surf the Internet at the same time.

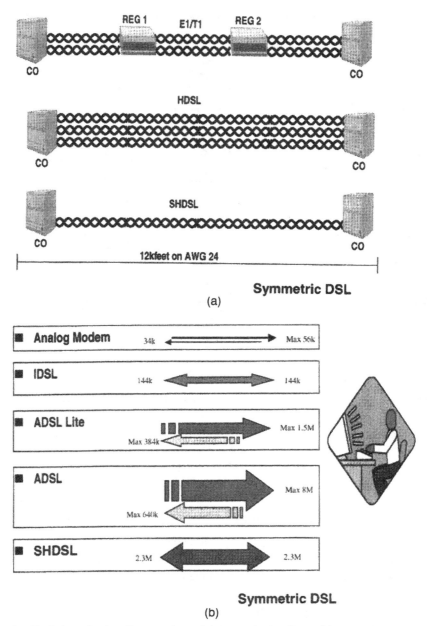

Figure 7-2 Evolution of subscriber services: (a) transmission lines, (b) transmission speed.

The greatest application of xDSL, in terms of economics, is for multiple-tenant units (MTUs), such as a high rise or a large apartment complex with many tenants. Currently such premises must use individual leased lines as shown in Figure 7-4. The switching point (SP) facil-

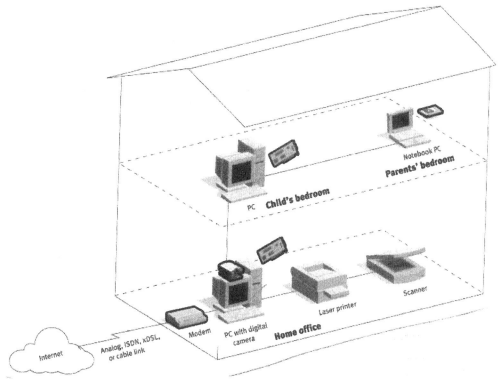

Figure 7-3 Phone, fax, and PC on a virtual PBX.

ity and the network access points are at different locations and are connected by high-speed connections, such as operational carrier level three (OC-3) or T1. The disadvantages of such a scheme are the total cost to the tenants and the fact that the building owner receives no compensation for the building wiring investment or its upkeep. The competitive local exchange carrier (CLEC) is responsible for the lines to the building and usually, in the case of Ethernet and T1 lines, those inside the building. This system, a carryover from telephony, is very inefficient for carrying data. An example of a typical building with tenants utilizing DSL technology being supplied by a CLEC via a local central office is shown in Figure 7-4. CLECs are proving the effectiveness and reliability of DSL by rolling out services at many locations.

The scenario shown in Figure 7-5 has the advantage of utilizing existing copper wiring. However, it has the economic disadvantages that each tenant pays for an individual Internet and the building owner is not remunerated for investment in building wiring.

A more economical scheme for tenants and the building owner is shown in Figure 7-6. Here a single or double twisted pair connects from the IP to a router that connects the user to the Internet. The connection of DSL lines to the premise are to the telephone service room or rack, which is usually owned by the premise landlord. The advantage of bringing DSL inside a campus or building is that it becomes a revenue-sharing opportunity for the building proprietor. The

With Individual Leased Lines...

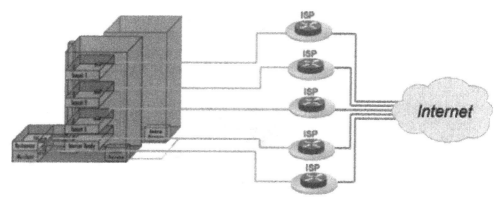

Figure 7-4 An MTU with individual leased lines. (*Source:* Courtesy of Interspeed)

With CLEC xDSL Lines...

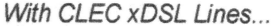

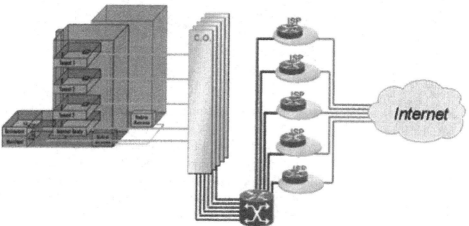

Figure 7-5 An MTU utilizing a CLEC with xDSL lines.(*Source:* Courtesy of Interspeed)

copper lines installed when the building was built or remodeled become an income-generating asset. This allows the landlord to negotiate with the local exchange carrier (LEC) and the tenants for Internet services.

The building landlord may be approached by the local ISP or approach the ISP to provide DSL service to the building. The price of the lease may be on a per-line or per-building basis. The ISP and landlord must decide on electrical usage and location of the *digital subscriber line access multiplier* (DSLAM) usually located where the copper wires terminate, in the basement.

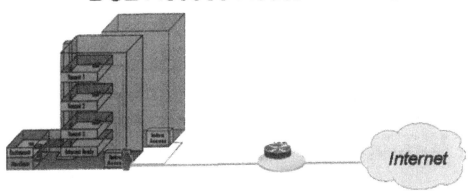

Figure 7-6 An MTU connection using a DSL access router. (*Source:* Courtesy of Interspeed)

7.4 DSL ACCESS EQUIPMENT

There are many solutions to deploy DSL under the category called DSLAM. Figure 7-5 illustrates a DSL connection employing a DSLAM. All lines from the CPEs are terminated at the DSLAM and then sent individually to the ISP. This requires the ISP to maintain information pertinent to each carrier point of entry (CPE) and to assume responsibility for performance and efficiency issues. Not only does this method impact the method by which the ISP provides service, but it also uses the leased lines between the building and the ISP inefficiently.

Figure 7-6 illustrates a potential connection of DSL using a *digital subscriber line access router* (DSLAR). In this scheme, all lines are routed to the ISP via the gateway router so that all intelligence necessary to maintain efficient transmission remains at the DSLAR.

7.5 SUMMARY

The primary reason that DSL has not reached its full potential is its limited reach and the fact that Internet servers are not capable of passing multiple high-speed DSL video streams because most mainstream routers are not designed to take thousands of connections on the same port. There are, however, many faster boxes arriving from numerous vendors that are designed to accomplish those functions. The disruptive effects of laying new cable—including hiring labor, rewiring utility poles, digging up streets, and disrupting office routines—make DSL the choice of many businesses. Since the copper is already available, the disruption of a DSL solution is negligible.

Selecting the correct xDSL for business or residential application depends, to a large degree, on the type of DSL services provided in the area. When several services are available, a careful study should be conducted to determine the present and future needs of the organization for data speed, number of users, and cost. When these factors are equal from more than one vendor, the choice should be made based on the service provider's track record for service and dependability.

The DSL providers that emerge as winners will be companies with the most efficient cost model for routing IP packets and the greatest QoS and variable cost structure. They will be differentiated by cost, access to services, and speed of transport.

Several of the deep-pocket communication players are getting into the DSL game, which will escalate DSL offerings. However, don't count out the cable industry with broadband access. For example, Cisco and Excite@Home showcased a "wired home" with high-speed cable modem service. The home includes a wireless network, an Internet-connected fridge, and automated software that allows setting the home security system and operating appliances.

The myriad new network offerings will provide new and economical services to the consumer.

QUESTIONS

1. What is the definition of DSL?
2. What is xDSL?
3. What are the most probable applications of DSL?
4. What are the advantages of DSL over T1 lines?
5. What troubleshooting techniques are adaptable to DSL lines?

Planning the Wiring Installation

8.1 INTRODUCTION

Planning a wiring system is the responsibility of the person designated as the telecommunication communication manager or data communication manager. In this chapter we will refer to that person as the **TCM**. The responsibility of the TCM with respect to the wiring system is to assemble the team that will design, evaluate, install, test, and maintain the wiring system that will be utilized in the telecommunication network of an organization. The team should consist of everyone who will be involved in the project. The list may include any or all of the following, some of whom may only supply information:

- The **architect** who designed the original building or who will design the new building, if the installation is a new construction.
- The **managers** of the departments that will be using the network. They must identify the number of telephones, PCs, printers, plotters, etc. that are needed in the final installation. If the installation is to connect to an old system already in place, they must identify the types and number of presently operating equipment. It is also very important that each manager identify future needs.
- A **data communication engineer**, who will evaluate the final inventory of equipment and design a LAN that will operate the final system. It is this person's job to select an operating system, the interfacing equipment, and the software to control the LAN.
- The **facilities engineer**, who will determine the location of equipment rooms and their suitability; the location and size of equipment closets; the location of existing cables and if those cables can be utilized for all or part of the installation; the existence of cable termination equipment or other equipment that must be removed; and the location of all cabling

ducts, open conduit, tunnels, raised flooring, and any other areas that might be utilized for cable installation.

- **Technical support personnel**, who will be performing the wiring installation, system testing, and/or maintenance. If these people do not exist within the company, someone must be hired or suitable personnel must be sent to school to train on these subjects.
- **Outside consultant** for telephone and/or data communication. Consultants can be hired for any part of the LAN installation, including system design, wiring layout, wire installation, wire testing, equipment installation, equipment installation and testing, and system testing. The maintenance of the system can also be contracted to an outside firm.

The TCM should become as knowledgeable as possible about the new network and every step of the design, installation, and testing before the team is assembled for the first meeting. This is especially important if some or all of the project is to be accomplished by outside contractors.

As we noted earlier in the text, it is important that planning and installation guidelines be established and adhered to whenever any major system needs to be installed, upgraded, or removed. Such guidelines should include cabling, services, and selection and installation of devices. The following sections present some of the most important aspects that must be addressed and the reasons behind them. A checklist of the items should be developed and followed.

8.2 PROJECT SCOPE

The scope of the cabling project should be determined before the project begins. If the project team is fortunate enough to be dealing with new construction, the work will not be complicated by in-place cable and/or wiring systems. However, the items discussed in the following sections should be considered and the guidelines should be followed for a new project, modification to an existing system, or addition to an existing system.

8.3 EXISTING CABLING

All existing cable and wire runs must be well documented in any cabling system. If the project requires a new cable installation in a facility where there are existing cable and wire systems, the existing systems must be well documented in advance of any work. Such documentation should adhere to the company's cable tracking system guidelines. Plans should be available showing all cable ways, under and over cable troughs, risers, and so forth, *including all unauthorized but existing cable runs*.

If the building has more than one floor, a layout of each "service floor area" is required. Each plan should include power, plumbing, and HVAC (heating, venting, air conditioning) systems. Total square footage should be on each plan per floor or area.

8.4 USER POPULATION

The projected number of users within the facility must be known to estimate the number of service ports and the type of service levels that will be required. This step will require input from the user community and the building and/or planning departments.

8.5 NUMBER AND TYPE OF WORK AREAS

The current number and type of existing work areas must be documented. This should include the number of work spaces, office spaces, classrooms, data centers, lab areas, test and engineering areas, and support areas. It is good practice to determine if the facility work areas are to undergo any structural changes, such as office consolidations, work space doubling, and/or work area conversions. Any changes will affect four areas:

1. The potential user population
2. Number and type of work areas
3. Number of service levels required
4. Type of system support required for each work space

8.6 DOCUMENTATION AND ROOM-LAYOUT DATABASE

A physical layout of each work space and/or work area should be maintained on a database. The layout should include the following:

1. Inside dimensions.
2. Placement and size of windows, doors, pillars, etc.
3. Location and type of all air conditioning power outlets.
4. Location and corporate identification number of all room cables (copper media, fiber, combined cable systems).
5. Location and outlet type of all telephone equipment. This should include all modular jacks, station wiring boxes, and cable system/telephone outlets. This type of documentation is an excellent application for a CAD (computer aided design) system. CAD is a computer graphics program on which the designer can create, modify, and display engineering drawings and diagrams. Typically, such programs are menu driven and allow the designer to rotate any part of the drawing, zoom into a specific area for close detail examination, or scale back to view the entire diagram.

A two-dimensional drawing of a typical office showing various characteristics of the office layout is shown in Figure 8-1. The layout is best drawn on a CAD program such Tur-boCad [R].

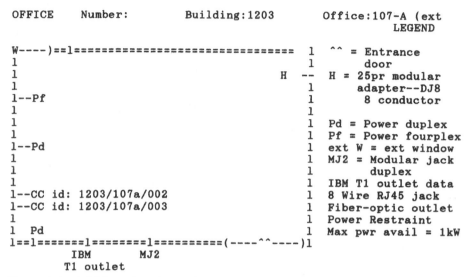

```
OFFICE     Number:          Building:1203      Office:107-A (ext
                                                        LEGEND

W----)==1================================= 1   ^^ = Entrance
1                                          1        door
1                                       H  --  H = 25pr modular
1                                          1        adapter--DJ8
1--Pf                                      1        8 conductor
1                                          1
1                                          1   Pd = Power duplex
1                                          1   Pf = Power fourplex
1--Pd                                      1   ext W = ext window
1                                          1   MJ2 = Modular jack
1                                          1        duplex
1                                          1   IBM T1 outlet data
1--CC id: 1203/107a/002                    1   8 Wire RJ45 jack
1--CC id: 1203/107a/003                    1   Fiber-optic outlet
1                                          1   Power Restraint
1  Pd                                      1   Max pwr avail = 1kW
1==1=======1=======1=========(----^^----)1
        IBM        MJ2
      T1 outlet
```

Figure 8-1   A CAD-produced room layout.

8.7 NUMBER AND TYPE OF DEVICES REQUIRED

The number and type of dumb terminals, workstations, and word processors must be identified. Many of these devices require special port assignments, and some will need specific controller customization. With the device type, a planner can anticipate the number and type of controller ports that the user community needs.

8.8 PHONE AND/OR STATION EQUIPMENT REQUIREMENTS

The number and type of phones that will be required for each work area must be determined. Presume that each area will need at least one phone if the information is unavailable at the time of inquiry.

8.9 MAXIMUM POWER ALLOCATION AND THE NUMBER OF POWER OUTLETS

Power consumption is an important issue for building managers. Even in this age of low-power-consumption devices, a maximum power specification needs to be determined for each area, enforced, and periodically checked to ensure user compliance. The planning group should be able to anticipate the users' product and device mix to estimate power requirements and ensure safe power margins.

Historically users have requested and installed more devices over time and with changing technology. As this occurs the number of power connections can exceed the number of outlets within a given space. The users generally resort to power strips or extension cords with multiple outlet ends. This usually leads to a "power crunch." To better manage AC power loading and user tendencies, the organization should do the following in all work areas:

1. Ban the use of power strips.
2. Ban all extension cords.
3. Ban the use of plug-in multiple outlets.

If at all possible, plan to use 80% of power availability at most. This will ensure a 20% safety margin. This 20% safety margin will probably be consumed as users attach unauthorized devices in their office or work space areas. These devices will include pencil sharpeners, answering machines, air purifiers, radios, stereos, TVs, and so on. Such noncompliance is the norm. Expect it, plan for it, and budget for it. The number of outlets will determine the maximum number of electrical/electronic devices within each area. An insufficient number of outlets will force even the most dutiful employee to install power strips.

8.10 TEST EQUIPMENT AND COMMITMENT TO SUPPORT PERSONNEL TRAINING

Ensure that support personnel have the appropriate equipment to perform normal and necessary testing, installation, and repair functions. Management must have a continuing commitment to training.

Training is a major issue among support departments. All personnel need to be fully trained and checked out on the equipment required to do their jobs. As new products are introduced into the system (be it a large data facility or a small LAN network) training must be updated continually.

8.11 TELECO, VOICE, AND DATA SUPPORT ROOM REQUIREMENTS

The following is a partial list of some of the equipment that may be needed by the various work areas to provide adequate support services. Such areas would include all teleco telephone switching rooms, patch panel/distribution areas, equipment rooms, and other facility support rooms.

1. *Telephone.* Preferably a two-line teleco circuit. In this way, a tech or support person can give assistance to the user community and still have an open line for additional support or query. A two-line circuit will also provide a communications link between the support room and other service centers that may be needed for problem determination and diagnosis. A third line should be considered for DSL.
2. *Online workstation.* This PC should have all levels of current and past communications programs on its hard drive. Since this is a test station, it should include all terminal emulators and control programs that are used in the facility to attach to the host. The workstation should have multisession capability. This feature will allow multiple sign-ons to assist in

problem diagnosis. One session could be a host attachment, a second session would maintain contact with a service or problem-history log, a third session could be used for online diagnostic support service, and a fourth session could be used for online messages.

3. *Network diagnostic tool.* This could be a software package loaded into the workstation but it might also be a separate network monitor. This will give the support personnel the ability to monitor current line conditions and status and will assist in the diagnostic efforts to solve a voice or data problem.

4. *Voice and data test equipment.* The importance of test equipment and the proper training to use it cannot be overstated. The following is a list of some of the test equipment that support personnel can use in maintaining the data/voice facility.

1. Manual tools: Universal tool kit

2. Cable and wire equipment

   Time domain reflectometer, TDR (cable break checker)

   Inductive amplifier and tone set

   Twist-on (one-piece) RG coax connectors

   Crimp tools and connector parts

   Wire strippers and cutters

   Coaxial cable strippers

   Wire-wrap tools (manual and electric)

   Insertion and extraction tool

   Ribbon connector crimp tool

   25-pin gender changer (to interconnect two data cables)

3. Fiber equipment

   Fiber-optic power meter

   Optical continuity checker

   Fiber identifier

   Optical TDR

   Local splice alignment and measurement set (LSAM)

   Emergency optical splice kit

   Fusion splicer

4. Signal and voltage test equipment

   Continuity and circuit tester

   Current meter

   Digital multimeter

   Oscilloscope

5. Telephone test equipment

   Inductive amplifier and tone set

   Punch-down tool (impact tool)

   Wire-wrap guns and supplies

Line test set

Modapt (8-wire modular plug-ended testing adapter)

Modular crimp tool and modular plugs

6. Network diagnosis hardware

Breakout box (RS232 diagnostic tool)

Bit error rate test (BERT)

Modem test set (combo breakout box and bit tester for modems)

Protocol analyzer (monitor or simulation mode)

Pattern generators

Line monitor

7. Electronic repair

Solder and desolder stations

Integrated circuit (IC) module insertion tool and adapters

Variable power supply

Oscilloscope

Signal generator

Figure 8-2 depicts a well-equipped equipment room with many of the aforementioned devices.

8.12 ENVIRONMENTAL CONCERNS

All premise cable and wire products are expected to function normally in a typical facility environment. Two of these environment parameters are the temperature and humidity variations.

8.12.1 Premise Cable and Wire Group

For cables of all types the acceptable temperature range is –40 to +80 degrees Celsius (–40 to +176 degrees Fahrenheit). There is no humidity specification.

8.12.2 Teleco and Support Rooms

The least temperature-tolerant device will set the range for the entire room. Normal operating temperature and humidity ranges will be found in the operating specifications portion of the user's manual for each device.

8.12.3 Air Conditioning, Dust Removal, and Chilled-Water Supply Requirements

In most facilities, air treatment is required to maintain acceptable levels of temperature and humidity for the proper operation of installed machines within a computer floor or data center. Ensure that the air treatment plant is of sufficient capacity to maintain conditions within acceptable limits for current and near-future expansion plans.

Equipment Room

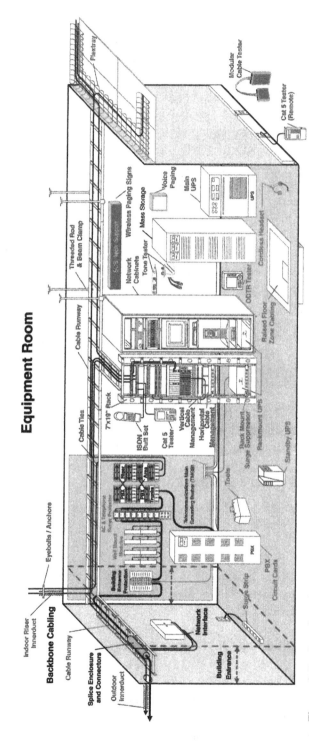

Figure 8-2 A well-equipped equipment room. (*Source:* Courtesy of Anixter Inc.)

The planning should incorporate forecasts of future device configurations and workloads. A plan that is able to meet requirements for the next two years should be considered acceptable. A 5-year plan is better, and if the budget and facilities will handle it, a 10-year plan is best.

Due to certain applications, especially in the DASD (direct access storage device) and thin film build areas, specifications are generated concerning the air quality in a work space. In such environments, dust and/or particulate matter in the air is monitored and, if necessary, filtering systems are employed to maintain the pollutant count in the air within certain prescribed levels. If warranted, the lab facility may need separate air cleaners, scrubbers, and filters to meet required specifications.

Another aspect of the facility that must be addressed is the chilled-water supply for large CPU (central processing unit) systems, current and proposed. Some CPUs (e.g., IBM 308x and 309x systems) will need chilled-water service for temperature maintenance.

8.12.4 Plenum and Non-Plenum Cable Applications

Plenum cable is cable that has met the low flame and low smoke characteristics as required by the Underwriters Laboratory (UL) to be used for installation in air-handling plenums (ducts), without conduit in ducts, voids, and other spaces used for environmental air. These cables conform to the NEC articles 725-2(b) and 800-3(b). Each cable is comprised of insulated copper conductors often surrounded with teflon or mylar to give the low flame and low smoke characteristics.

8.13 GROUNDING AND BONDING

As was stated in Chapter 1, electrical grounding of equipment and devices was originally intended to be a safety measure to prevent electrical shock to personnel. However, modern grounding and bonding systems are being designed to provide a low-impedance path for noise and voltage transient protection. Such electrical events could disturb signals in communications and electronic devices.

The following is a list of some of the grounding areas to be considered:

Structural ground
Signal and data path ground
Lightning protection
Existing plumbing grounds
Grounding equipment racks
Grounding racks, panels, and cable shields
Measurement of building ground potential difference and ground path resistance

8.14 CABLE NETWORK MECHANICAL SUPPORTS

Cable network mechanical supports are comprised of cabling passageways, channels, trays, and distribution racks and will be discussed in detail in Chapter 9.

8.15 ELECTROMAGNETIC INTERFERENCE

There are four types of electromagnetic interference:

1. Electrostatic discharge interference (ESD)
2. Induced electromagnetic interference (EMI)
3. Low-frequency interference (LFI)
4. Radiated frequency interference (RFI)

These factors were covered in Chapters 2 through 4.

8.16 USER DEVICE OWNERSHIP

A big question for each organization is, "Who will be responsible for device ownership?" The answer will depend on the position the organization takes concerning ownership of users' devices. Before this decision is finalized, the following question should be answered: Will the work space device(s) be owned and therefore tracked and maintained at the department level? If the answer is yes, then management will be responsible for inventorying and expensing the department's work space devices. This allows each department to operate as a business unit. Each department will manage its own costs and expenses. Business justifications will have to be made to upgrade and augment current device hardware or computing facility on a departmental as well as a specific employee basis. If the answer is no, then another department (other than the user's department) will inventory, maintain, and track all device hardware. This department, for example, may be in charge of all support hardware devices throughout the function as well as the computer center support services.

Departmental device ownership has certain clear advantages over other ownership schemes, including ease of support. However the down side is that the user community may believe it is at a disadvantage due to "packaged product and service assignments" generally believed to meet user requirements. Some users may want to be able to choose their equipment instead of having it assigned. In either case, an inventory must be maintained and updated on a regular basis.

8.17 HOT HOST SERVICE

Another question worth considering is whether or not the company will provide at least one hot service in each work space with a known connectivity path. In this approach the telecommunications and support departments will provide at least one hot or active line per office that will provide a standard connection for a range of currently used workstations and/or PCs. Any devices outside the specified norm will have to be called into the service desk for special port assignment or software assignments. If the facility has enough controller or mux (multiplexer) ports available, this is something to consider.

The advantages of this system are as follows:

1. During initial space occupancy by users, there will already be a defined system service available in each work area.
2. During moves, adds, or changes to the user population or system services, the hot service concept will greatly decrease system service downtime.

This approach will assist the strategy, planning, and support departments in driving toward a fully complemented information age office. In this office, all hardware and software services are already in place prior to use. The connectivity paths allow almost any host-to-user path and provide standalone workstation capabilities.

Standalone workstations allow the user to detach from the host to do standalone processing. This separate activity frees the host for other work. This system works well in a facility that is fully populated or in which there is a high turnover rate in the user community.

The disadvantages are obvious. It is a resource hog. If the facility has 200 offices, at least 200 current-level services need to be provided. This application commits to service 200 ports. If by chance the facility is not fully populated, many ports remain hot and not connected.

8.18 BUILDING OFFICE AND DEVICE INVENTORY

A building office and device inventory, to be complete, must include the following:

1. **Building and room location.** Each room must be documented by corporate building, floor, and room identifier. This documentation should include the listing of the current number of power outlets in each work space on the room layout diagrams.
2. **Physical device hardware.** This includes the type and model of each workstation device. The following example of an IBM-only shop will illustrate some of the detail needed for the documentation. The responsible group must indicate the type of terminal in each office space.

Dumb terminals
Dumb terminals do not have internal memory and therefore cannot do any editing without referencing back to the host or controller.

Smart terminals
Smart terminals are those that include memory. This category includes PCs and workstations.

PRINTERS
Remote:
  Line address:_____ Port address:_____
  System:_____ CPU or LAN attached:_____
Local:
  Line address:_____ Port address:_____
  System:_____ CPU or LAN attached:_____

Laser:

Line address:_____ Port address:_____

System:_____ CPU or LAN attached:_____

PLOTTERS

Remote:

Line address:_____ Port address:_____

System:_____ CPU or LAN attached:_____

Local:

Line address:_____ Port address:_____

System:_____ CPU or LAN attached:_____

FAX

Remote:

Line address:_____ Port address:_____

System:_____ CPU or LAN attached:_____

Local

Line address:_____ Port address:_____

System:_____ CPU or LAN attached:_____

3. **Configurations.** Any special PC, workstation, or device configurations should also be noted in the inventory.

4. **Communications software and release level.** Currently installed communications software and microcode levels must be noted. Make sure that equipment and software meet current standards; to this end, a physical inventory of installed inventory is necessary and must be checked against current-level hardware and software requirements. This will help keep the user-device community updated and avoid incompatibility problems that can develop as newer levels of microcode, terminal emulators, and/or hardware improvements are introduced into the system.

Example: SYS2 Emulator xx.1 through xx.5

Example: Please note that xx.3 requires a patch to work on system DC12. This patch is #445 and is available from department JJ15.

5. **Room cabling information.** The total number of connecting media within the room or office space must be noted. Special attention must be paid to how many of the office cabling and or individual coaxes are utilized and how many are not in use. These numbers become important to office/space planners as well as department and functional groups. This information can be used to arrange the user population around specific needs and services if the data are available to everyone during the planning phase of a corporate move. Dedicated wiring and/or the presence of a corporate cabling system must be noted by the specified wiring label scheme. Finally, note which cables are connected to which devices.

6. **Telephone services.** The number and type of phone outlets by room location and all trunk and line routing information must be noted.

7. **Controller requirements.** The controller that is required or specified for each terminal service along with the controller type, serial number, configuration, and microcode level must be noted. Also indicate where controller lines are terminated. A back-up microcode diskette should be maintained at all times.

8. **CPU and/or distributed services, including current status.** CPU services that are requested or required should be examined. It may be that CPU services can be provided through front-end processors or concentrators. Maintain a running inventory of network monitor screens or concentrator screens. These screen displays should indicate current available status. That is, there should be an information panel that indicates which services are provided to the user community and which systems and services are currently up and running as well as which systems are down. For those systems listed as being down, an estimated time to be back up should be given.

8.19 NETWORK AND PLAN DOCUMENTATION

Network and plan documentation requires documentation of the logical network configuration starting at the end user devices through all connections back to the servicing centers (this could be a CPU or a concentrator). This requires the following steps:

1. Documentation of all software layers for full implementation of end user available services.
2. Recording of all available software systems.
3. Recording of all subsystems.
4. Documentation of the physical network topology.

These items are accomplished by showing all box connections on a two-axis coordinate system or corporate approved location scheme for each area on the drawing. (As a suggestion, have cards or placards made up and positioned in clearly visible locations around each floor and work area, with large building coordinates alphanumeric. This can be a big help in locating something on the floor or in the area. Too often this numbering scheme is placed on equipment racks or on support posts not visible from all vantage points within the area.)

There are two methods to accomplish this task. One is to use the tried-and-true manual method, which is very labor intensive. The other is to use a CAD program. This approach is computer power intensive.

In either case the drawing should include all areas on the floor plan, including power, HVAC venting, work areas, patch panels, equipment rooms, and wiring/service closets, with dimensions, power, and access indicated on plan.

With the manual method all drawings, other than the floor plan, should be done on clear vinyl or plastic so they can be placed on top of existing floor plans or stacked two or more on top of each other to show various components of the network.

With a CAD system, the rooms and work spaces are displayed in two or three dimensions with various facility service layers superimposed on the base drawing. (The designer can use a specific color for each group or service layer.) In this way a graphic with several layers superimposed on it can be more easily understood by viewers. These would include the following:

1. HVAC system under and above the data center floor.
2. Power, including locations of outlets, types of service, and service extensions under raised floor areas and above ceiling locations.
3. Cables, CPUs, DASD strings, controllers, telecommunications racks and equipment, furniture, and chilled-water layout.
4. All cable trays, pathways, and cable channels. This drawing should include
   1. All vertical and riser cable paths
   2. All access holes, between floor riser pipes, and through firewall holes (approved and otherwise)
   3. Dimensions of all sheetmetal trays and boxes
   4. Dimensions and measurements of sheetmetal trays and boxes above floor and below raised floor areas
   5. Radius of all tray turns or elevations
   6. Support and suspension hardware for all sheetmetal or plastic trays
       If cable trays and cableways are being ordered for installation, try to get them large enough to accommodate a long-term plan. Factors to consider when planning are as follows:
   1. Tray capacity
   2. Suspension type and weight capacity
   3. Access points
   4. Once installed, cable trays above dropped ceiling panels and cableways under raised computer floors are seldom if ever enlarged.
5. All cable wiring routing, including voice, data, video, HDTV, and fiber optic. This drawing should show all lengths of runs, drops, and splices. The individual splices may be difficult to show, but if the cable line is important a special detail layer can be added. Any unauthorized cable runs or cable runs that violate city building codes or corporate guidelines should be included.
       An action plan will need to be developed to correct any violations of existing codes and procedures. Be sure someone is responsible for tracking and closing out all outstanding infractions. If the city or local building department is involved, be sure all work is appropriately signed off.
6. Telephone circuits. This plan should include the following:
   1. Bulk cable routing into teleco rooms
   2. Fanouts into 66 termination punch-down blocks
   3. All PBX terminations

7. A localized drawing can be done showing specific cabling detail. For example, a template can be made up to show a specific wire or cable system routing, through patch and wire closets and final terminations.

8. Each floor plan should show alignment points so that one floor plan can be superimposed over another floor of the building. This will allow a planner to work effectively with an isometric building or floor plan for cable measurements.

9. Standard architectural and computer symbols should be utilized, with all special characters noted in the plan legend.

10. Accurate network documentation is vital for both planning purposes as well as problem diagnosis. With a complete guide to the facility cabling system, a better understanding can be achieved and more efficient work can be accomplished.

8.20 QUALITY AND ELECTRONIC CONTROL

Ensure that someone verifies all cable runs for actual distance run from beginning to end. This should be done for all existing cable runs to get a cable distance baseline for each facility.

Each type of cabling has a maximum length standard (with or without repeater or relays) as set by the manufacturer. If the cabling is part of a system, check with the system or design specifications index for maximum length statistics. Ensure that all length, splice, and signal strength criteria are meet.

8.21 SERVICE IMPACT SEVERITY CLASSIFICATIONS

All work that affects any system or service needs to be categorized by impact severity classification. This classification must be determined by the work or task planner. Be sure to be include the category of impact in all correspondence to all department heads. Impact should also be a bulleted item on all presentations. Impact categories are as follows:

Category 1 Impact: No risk. No impact on services or systems.

Category 2 Impact: Low risk. No noticeable impact on normal services.

Category 3 Impact: Moderate risk. Small chance of service hits occurring. Minimum exposure can be handled by the installing and support crew.

Category 4 Impact: High risk. Good probability that there will be some service interruptions due to work activity scheduled. Support staffing required. This level of impact requires one or two levels of management to install a power connector device if the work space cannot meet minimum attachment needs.

Category 5 Impact: Very high risk. Definite service impact due to the nature of work activity. Additional field and support personnel required for installation. System restoration will require three levels of management signoff and facility management signoff.

When doing a physical inventory of office equipment, make note of all devices attached in the area. It is important that the total power load on the electrical system for the facility be known and an advisory notice to management be updated on current and projected power usage requirements.

Some facilities will have a power crunch because the initial planners never foresaw a time when users might require two or more terminals/PCs in each office. Advance planning will ensure an adequate power safety margin over the long-term plan.

Historically, the first office-installed devices were dumb terminals, which needed only one AC outlet for power. However, with the advent of the PC many things changed, including the number of individual products that required their own outlets.

In a modern office space planners should install six to eight power outlets due to the power requirements of current workstation systems. The following is a minimum hardware configuration for an office setup requiring a six-plex power outlet:

Outlet	Device
1	CPU
2	Display
3	Printer
4	Mouse (some require separate AC power)
5	Fax, laser printer, expansion chassis, extra display, external modem, additional PC, desk light, radio, TV, etc.

8.22 SERVICE AREAS AFFECTED BY WORK TO BE PERFORMED

It must determined in advance which functions and areas will be affected by cable installations. If power, telephone, or existing computer services must be interrupted to implement the installation plan, all affected areas must be notified of the date and time of the service outage. Two weeks' advance notice is a good lead time for announcements.

All pertinent information regarding service interruption and contact phone numbers must be put on all online message systems. In this way anyone logging onto a host system will get the message and know whom to contact. A contact point and meeting schedule should be set up well in advance to discuss the service interruptions with any and all interested groups.

These timely advisory meetings in advance of the actual work will assist in minimizing the impact of the outages on the various departments and functions within the facility.

A rating of impact severity levels by work activity must be established. These ratings must be included in all advance documentation and meetings concerning the work and any outages because of it. All department heads within the facility must be copied on all correspondence of the upcoming meetings. Include the schedule of meeting times and locations as well as a contact name and number within your organization for additional support. Post this notice within each facility on a general news bulletin board at least two weeks in advance of the actual work.

Finally, the "company contact" of contractors and vendors that are to be working within the facility must be notified of the advisory meeting dates, times, and locations.

8.23 REVIEWING BUILDING PLANS AND CABLE REQUIREMENTS

The following steps are necessary:

1. Copies of the floor plan for all affected areas should be available to your organization.
2. Telephone, voice, and data cableways should be documented and plans made available. These plans should include all existing cable trays, troughs, and overhead and below raised floor areas.
3. Pay special attention to all cable/power/coax runs that are not documented elsewhere but are known to exist. It is important that these cable/power/coax runs are documented for future reference. It may be that they are not to current local or corporate wiring/cabling/safety codes. If that is the case, then notify department planners to write a plan to correct any violations of current codes.
4. The location and function of all wiring closets must be documented. If your facility has dedicated communications rooms for wire/cable service as well as telephone equipment rooms, determine current service level capacities. It is important to determine the space utilization for future planning.
   Determine if the support rooms have the floor or wall space to support more hardware. These support rooms could be corporate teleco rooms (telephone equipment rooms) as well as wire/cable service rooms. Perhaps with some form of space efficiency stacking racks or trays, the same room will accommodate much more hardware and/or patch/LAN/concentrator racks.
5. If your facility has combined telephone and data facilities closets, and if your planning organization expects the facility to grow, then it is advisable to separate teleco and data into dedicated rooms.

Note that in any facility the emphasis is usually on the user community requirements and not support facility needs. Therefore, space requests for support areas may be met with substantial opposition. You may have to fight for it. Better now than later once all the space has been allocated according to long-term goals and service-level commitments.

8.24 SERVICE DESK

A service desk should be established to track the health of the telecommunications system. To assist the service desk personnel, a tracking facility (file system or database management system) should be established that is available during normal working hours. Everyone in the service facility should be checked out on its proper use. Service desk job requirements should include a technical background and diagnostic skills. The person who maintains the service desk should be able to

1. Derive from callers all pertinent information concerning actual problems.
2. Call up system support facilities to determine if there is an actual problem or a user problem.
3. Begin first-level diagnostics once it is determined that an actual problem exists. This will determine if the problem exists in the user's equipment, line or cable service, or system service.
4. Initiate a problem report, send it to the appropriate department or organization, and track it.
5. Categorize incoming problem calls according to a number priority classification system. As an example, the problem classes could range from 1 to 10 (1 being the least severe and 10 being a major system crash with widespread service interruption throughout the organization). If, for example, a user calls up and indicates his workstation is down and a bit of checking shows the system and other personnel within the same department are up and running, the problem class would be a 1.
6. Assign the proper telecommunications or service department's time-to-repair period according to problem class. The repair time by problem class should be discussed with all appropriate support service groups and agreed upon. Within an organization, procedures should be developed to handle various types of problem calls.
7. Develop and adhere to an escalation procedure concerning the repair time window. If for some reason the time-to-repair exceeds the prescribed amount for a certain level of problem, the next level on the escalation process should be invoked. That may be additional support personnel and/or managers being brought into the picture. All repair activities should be monitored by the service desk. Each higher level represents support personnel with greater knowledge and technical expertise in the area. Management should be notified once a service call goes beyond the first service level.
8. Ensure that all parts are ordered. This includes required parts, spare parts, and cabling systems.
9. Establish quality control for all cabling runs, service terminations, and device-to-service attachments.
10. Ensure that each member of the installation, repair, and update team has had sufficient training to perform all aspects of the work. This requirement includes all parts, testing, and installation equipment as well.
11. Establish overall target dates for each project. Break each project down into subsections. Set time frames for each work subsection. Schedule status meetings prior to subsection work completion. At these meetings discuss the current status of work and any problems with assigned activities or support services. Review with all work groups, corporate and vendor, level progress and overall project completion targets.
12. Maintain a clear management path for status and problem reporting by establishing a three- or four-level scenario concerning overall project completion. Anticipate problem areas and bottlenecks. Try to resolve these issues beforehand. Discuss impact levels on

related systems and tasks, from the best scenario of work completed on time to the worst-case scenario.

13. Maintain constant contact at the service desk, whose personnel will be responsible for updating the tracking facility with the work status. The tracking facility should have the capability to issue reports on current work status. These reports will include such items as subsection work completed, estimated completion time by subsection work assignments, work delays, and overall work progress.

8.25 SCHEDULING THE JOB

The TCM and his or her staff, after considering the aforementioned details, must develop a plan that will schedule every step of the design and installation of the system. This includes the following:

- Communicating with everyone involved in the project, both company personnel and outside consultants, contractors, and material and equipment suppliers.
- Establishing the installation schedule with specific benchmarks or steps showing progress of the project.
- Coordinating contractors and vendors in the installation schedule.
- Scheduling equipment and system performance evaluation test periods.
- Determining the ramifications on the day-to-day operation of each department that is to be affected in an occupied area and scheduling installation work for the least disruptive times. Considerations must be given to moving a group to temporary quarters during installation. Where installation can be performed within the work environment, the TCM must consider the effect that the installation will have on the utilization of existing equipment. When an in-place system is to be upgraded, the primary consideration is downtime and the productivity of the unit. The TCM must evaluate carefully the cost of downtime and the installation labor cost of performing the installation during afterhours and on weekends.

8.26 WRITING THE REQUEST FOR BID PROPOSAL

The procedure for writing a request for proposal (RFP) is the responsibility of the TCM but may be assigned to a knowledgeable person. The task must never be requested of a vendor or contractor who is to respond to the request.

The proposal request must contain the exact specification of the work to be performed and the equipment that is required. The request for proposal should be clearly written to avoid ambiguity and vague requirements. The writing must be clear and concise so that there is no room for misinterpretation. Ambiguities in the document may generate different responses from vendors regarding the task at hand.

Prior to preparing the RFP the writer must know exactly what it is that is being requested. This will usually result in a great deal of research time. An RFP should never be prepared under the pressure of a deadline. Whenever specific products or equipment are required the author must specify the manufacturer and type. The important thing is to be specific. For example,

1. Install 20 four-conductor twisted-pair 22-gauge, plenum-rated cables in the locations shown on the attached blueprint.
2. Install 5 four-conductor stranded 24-gauge fire alarm and tray cables in the locations shown on the attached blueprint.

Most contractors will require a walk through the buildings in which the cabling is to be installed before making a bid. The proposal request should specify if the bid is to be lot or unit pricing. Unit pricing should usually be requested to show the cost of materials and labor for each item. In this way additions to the bid can be priced back to the unit price.

8.27 DOCUMENTATION RESPONSIBILITY

A single person should be given responsibility for system and wiring documentation. Documentation should be maintained on blueprint drawings, cable distribution logs, and key sheets. The *blueprint* is a detailed drawing of all the building space. In large facilities several blueprints may be required to represent the area. The *distribution log* contains a listing of each circuit in numerical order for quick reference. *Key sheets* are charts that detail every identifiable section of the communication circuit for quick reference in servicing or changing the system.

8.28 INSTALLING THE WIRING

The TCM should designate a single person to oversee the actual installation of the wiring and cabling to assure that scheduling is maintained and that no steps are omitted. This person must contact vendors at the appropriate time, assure their access to the facility, check the installation against the RFP, and sign off on the final acceptance test.

8.29 NEW BUILDING APPLICATION

If you are involved in designing a new building plan that includes a wiring system with a *cabling approach*, the following checklist will assist you in this task.

Installation Guidelines
Ensure that all communication cabling meets the National Electrical Code (NEC) guidelines and all local codes (some municipalities have additional requirements). The NEC covers all nature of cabling, such as plenum rated for air plenum environment, riser cables, strain relief minimums, lightning protection, proper fireball penetration routing and repair, and floor and ceiling penetration routing and repair. A portion of the code was repeated in Chapter 5. However, the latest code book should be consulted before the plans are complete. The NEC updates the code each year, and the latest NEC codes should be in the TCM's reference library.

8.30 ESTABLISHING A LABELING SCHEME

Establish a *cable labeling scheme* that includes the origination and destination data on cable labels and tags. The labeling scheme must also include the labeling of conduit. These tags should be on a nonsmear surface, easily read, and waterproof. Both a wraparound and an in-line type should be used. Also label all conduit. The labeling data may include building and room numbers as well as the sequence number for that particular room, lab, or work space.

Identify all rooms, labs, or work spaces, equipment closets, junction boxes or blocks, distribution boxes, and wall outlets with proper labeling.

Develop and maintain documentation describing all cable and optical fiber runs. This description should include point-to-point information, type of medium, overall length of medium, and passive and/or active components in the circuit run.

Document all hardware by creating a blueprint showing all patch panels, equipment racks, telecommunication closets, teleco closets, distribution panels, concentrators, multiplexer, controllers, and data circuit paths.

Establish and adhere to standard labeling for equipment and hardware. By having a standard labeling scheme for all cables and equipment, the technical team and engineers will be able to understand the layout and correct any errors. This will assure rapid turnaround on future changes and additions.

Blueprint all overhead raceways/trays and distribution breakouts. The same procedure should be followed for all under-floor distribution systems.

Establish the correct media termination procedures for each type of media to ensure technical competence in installing connectors, splices, terminations, etc. Establish a cable database that the tracking department can use to establish, maintain, and update the database for use by the technical staff, management, and end users.

8.31 DATABASE TRACKING SYSTEM

Developing a database tracking system before cabling a new facility or adding to or modifying an old facility seems to place the cart before the horse. The task will seem to be an impediment to "getting the job done" (the job being installing the cabling, connecting the hardware, and addressing the complaints of management and users). However, the person that is to be responsible for makings changes in the system, connecting new and additional equipment, maintaining and troubleshooting the cabling, and answering all management and user demands on the system should realize that a tracking system is essential. Chapter 12, on wire and conduit labeling, and Chapter 11, on database wiring management, can help in developing a plan that can be implemented as the conduit is installed, the wires are pulled, and the equipment is connected.

8.32 SAFETY

Safety, the last factor to be considered by most planning groups, comprises several categories (for example, the safety of the company employees, safety of the installers, safety of the plant, and safety of the equipment to be connected to the wiring). Here we will address the safety of the vendors and the equipment to be installed to the wiring system. The TCM or a designee should assure that the vendors are using safe procedures in the installation so that plant personnel and vendor workers are not in danger.

To protect the system users and the equipment that is being installed, the TCM must make sure that all local, federal, and IEEE codes are followed. Some of these are covered in Chapter 11 on installation of wiring systems. A current copy of the federal and local wiring codes should be part of the technical library and should be consulted to assure compliance with codes.

8.33 PLAN REVIEW BY ALL AFFECTED PARTIES

All plans must be reviewed by building and area representatives, and in contact meetings the purpose of the work must be specified. Affected departments are to be notified of actual work time, including prep time and system checkout.

Impact severity should be noted and understood by all departments. If the work requires system or service interruptions, make sure all department heads are notified well in advance so they can send a representative to planning meetings. At first glance this might not seem important, but departments will have personnel working at odd hours to take advantage of systems that are normally engaged during the prime shift.

Some work activities cannot be done during normal working hours; and for some departments, having personnel come in to start a back-up or to rerun a big job is not unusual.

A plan should be written and implemented to renew work when the activity is finished. This plan should provide for failure point checks and analysis if failures occur.

Back-up and technical support personnel should be in place during and after work activity to ensure that a smooth systemwide transition is made from beginning of work assignment to return of system.

If there will be a service interruption, by doing some of the prep work ahead of scheduled downtime, the time impact can be minimized. Whenever possible schedule all work (or as much as possible) on off-prime-shift hours to minimize impact of service interruption. Perhaps more than one work activity can be combined during normal service windows. This may take the combined efforts of more than one department, so make sure they can commit the support or work detail.

Instruct work teams to follow company cabling guidelines and perform frequent inspections. Once installation teams are aware of frequent inspections, the correct labeling practice will become part of the standard operating procedure.

Establish a service and information desk to assist with a smooth implementation of all work activities and plans as well as to serve as the first contact point for problem determination as noted in Section 8.24. Work procedures as well as problem determination procedures should be documented and understood by personnel who will have this job. The services provided should include guidance, consulting, and diagnosis. The service desk work assignment will need to be filled by a person who is trained in and familiar with all aspects of the facility, including system connectivity and user service as well as corporate long-term goals.

8.34 SUMMARY

The planning, installation, testing, troubleshooting, and documentation of a wiring system must be a team effort. Upper management must provide the resources and assign the most knowledge-able personnel to the development team. A team leader must be given full authority to hire expe-rienced team members and contract personnel necessary to ascertain the needs, design, evaluate, test, and certify the telecommunication system.

QUESTIONS

1. What team would comprise a planning group for a wiring installation in a soon to be con-structed building?
2. What team would be needed to provide wiring for a new department in an established facility?
3. When should outside consultants be hired to design a wiring system?
4. What is involved in the scheduling of cabling for a new office staff that is moving into an established facility?
5. Why is it important to write a request for proposal?
6. Who in your organization would be responsible for documenting a cabling project?
7. Identify, by example, the labeling scheme that you would use for new cabling that travels between two buildings, into an a equipment closet, through a cabling rack, and finally into a new office installation. Assume that once the cabling enters the building, half of the cables travel through the ceiling and half travel under the floor.

CHAPTER 9

Installing the Cable

9.1 INTRODUCTION

The installation of communication wiring is seldom simple. There are many things that should be considered before any cabling is installed. Ideally, the telecommunication communication manager and his or her team would design the communication wiring for the entire telecommunication system in conjunction with the architect's plans before the building was constructed. Then all the conduits, cable trays, vertical risers, outlet boxes, distribution closets, and distribution racks could be built into the building plans. This will seldom be possible as most facilities will have been utilized by the company, and cabling and equipment have been added as needed. New cables may have been added over older cables that were abandoned as equipment was updated and added.

A more realistic wish would be to have a plan of and to document all cabling and hardware. Most buildings or areas will have installations that were installed by several different people who have been transferred, or by outside contractors. Installations in the past were probably installed by each department on an as-needed basis with little or no coordination or overall planning.

Wiring systems seem to grow almost as if they were alive. It is estimated that the wiring collection in older companies is only 20% utilized. It is much easier and less expensive to install another cable than to find an open one.

9.2 MAKING THE PLAN

The communication manager or person responsible for making cabling changes and additions will have to take things as they are and proceed from there. If this job falls to you, you should do the following:

1. Obtain blueprints, room layout diagrams, wiring diagrams, and any other drawings that are available. A drawing should be made showing all known cabling and communication equipment.
2. Locate all closets, conduits, raceways, subpanels, cellular floors, ducts, and so on.
3. Identify each data communication station and the type of equipment that is located at that station.
4. Obtain database information if it is available.

Once all prints and information are assembled and the wiring manager or designee has determined the specific location and type of cabling to be installed, the TCM's team must asses the extent of the work and must determine who is responsible for the installation. It must be determined if in-house personnel have the time and expertise to perform the task or if the work will have to be contracted to an outside vendor.

The telecommunication plans should be based on data from the premise blueprints. However, the details are for telecommunication and not building structure. The details of power and cooling needs will require the assistance of plant environmental and power engineers after the telecommunication plan is completed. Of course, the power and cooling resources must be available, either in place or funded before the project can be concluded.

9.3 CABLE STRATEGY

Media networks can be physically configured in several different ways. Each has its own advantages and disadvantages. There are three basic connection networks that are used in most facilities. These are discussed in the following sections.

9.4 TWO-POINT CONNECTION STRATEGY

The most direct and simplest strategy in terms of wiring, connections, and routing is point-to-point connection (see Chapter 6). In this strategy all host services (controller port addresses) are represented on hardware connectivity panels. These panels are usually centralized into a physical point for the service host.

Room and lab cable runs are also terminated within close proximity to the controller port panels. With this method a physical connection can be made easily between host-side service and user-side devices.

An efficient application of a two-point strategy is premise cabling for a small building or office with a low move-add-change requirement and internally housed computer services. All host services could be confined to a single patch panel area while all room cable runs would be terminated within the same place.

The two-point cabling technique is the most direct interconnection between host and service user devices. Its disadvantages are as follows:

1. There will probably be a need to recable to meet changing needs.
2. Maximum cable length could be exceeded, causing material and labor cost to rise.
3. Propagation and signal-level problems could occur with long cable runs.

The decision to utilize this strategy must be considered carefully and with input regarding possible/probable future requirements.

9.5 THREE-POINT CONNECTION STRATEGY

The three-point strategy is less direct than the two-point method but is more flexible for both users and service requirements. With this method office/lab space cables are routed into designated local cable rooms, which function as remote cable distribution points for specific room connections (Figure 9-1). A room/lab in a designated area will have cable and media runs terminate on patch or connection panels in these local rooms. Cable feeder runs will be installed between the local rooms and the controller/host service patch panels located on or near the computer floor area.

The advantages of three-point connection are as follows:

1. Host service will be broken into several smaller links between host/controller service panels and the user's office or lab.
2. Tracking diagnostics, auditing, and line quality assurance will be made easier due to the segmenting of the cable runs.
3. Entire departments can be serviced by a centralized local room with dedicated host system connections. In certain instances there could be a security advantage to having sensitive host-to-user connections passing through one or two centralized points.
4. Individual cable runs are shorter, resulting in simpler upgrades and/or changes to multiple systems when they are available.
5. Service-level upgrades can be accomplished through one common distribution point for a group of office/lab spaces.
6. With the installation of feeder cables there should be enough cabling capacity to accommodate future user connections. In this case, running new cabling will become an uncommon event.

The disadvantages are as follows:

1. This system involves high cost for equipment and installation.
2. It requires adequate space and more facilities planning.
3. It will add to the overall complexity of line connections between host and the user and will necessitate an effective tracking system utilizing a cable management system.
4. The system will usually require a media database. This will require personnel to create and maintain the database and is another cost of providing this level of service.

Wiring Strategies

Examples of three-, four-, and six-point configurations

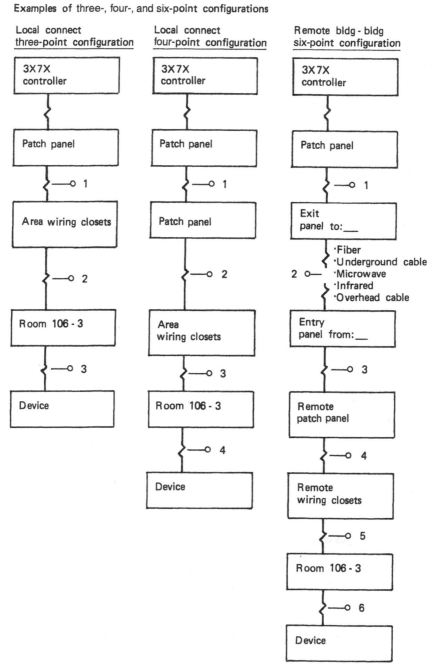

Figure 9-1   Examples of three-, four-, and six-point wiring strategies.

9.6 FOUR-POINT CONNECTION STRATEGY

This strategy embodies parts of both the two- and three-point strategies with a twist: Host services are provided within the local rooms.

A small service department computer is added to the three-point connectivity link. This CPU will be servicing a specific function or task group in a local area. Its breakout or distribution panels will be set up alongside the room cable connection panels. This dedicated host will be connected point to point in the two-point connection fashion and will allow users in that area the added advantage of locally attaching to a CPU for a specific application or program.

9.7 RULES FOR INSTALLING CABLE

All communication cable should be installed utilizing the National Electrical Code (NEC) guidelines discussed in Chapter 5. Communication cables are classified under the code to include fiber-optic cable, twisted-pair cable, and coaxial cable. The guidelines listed here were in effect in the 1998 code, reference articles 200 and 800. While most of the codes are rather stable, there are changes from year to year as new products are proven or research shows that excising codes may not offer proper safety to personnel and/or equipment. The most recent NEC code should be part of every technician's library.

9.7.1 Fiber-Optic Cable

Fiber-optic cables (Article 770 of the code) include the following:

- Nonconductive cables (those that have no metal members).
- Conductive cables, which contain noncarrying metal strength members.
- Hybrid cables, which contain optical fibers and current-carrying electrical conductors. These are classified as electrical cables in accordance with the type of electrical conductors in the cable.

Generally, fiber-optic cables are permitted to be included in current-carrying cables containing less than 600 V. The metallic carrying members of any fiber-optic cable should be grounded.

9.7.2 Copper Communication Cable

- Copper communication cable (NEC Article 800), such as twisted-pair and coaxial cables, should be supported at least 2 ft from power cables unless the power cables are enclosed in a conduit or raceway.
- Vertical riser cable made up of twisted-pair, coaxial, or fiber shall have a fire resistance to prevent the carrying of fire from floor to floor.

• Horizontal runs shall be made so that the possible spread of fire or combustion products will not be increased by the installation of the cable. Penetration through fire walls shall be blocked.

Only fire-resistant and low-smoke-producing characteristics shall be permitted to be installed in ducts and plenums or other space used for environmental air. Regulation information can be researched in Article 330-22 of the NEC, entitled Wiring in Ducts, Plenums, and Other All-Handling Spaces.

9.7.3   Outside Cable

For communication cable located on the same pole or run parallel to power cable, the following restrictions apply:

1. The communication cable shall be located below the power cable.
2. The communication cable shall not be attached to a cross arm that carries a power cable.
3. Climbing space through the communication cable shall be provided.
4. Supply service cable having less than 750 V running above or parallel to communication service drops shall have a minimum clearance of 12 in at any point in the span, including the point of attachment to the building.
5. An unground cable must maintain a separation of 40 in at the pole.
6. Communication cable passing over a roof shall maintain a minimum of 8-ft clearance above the flat roof of a building. The NEC gives an exception for other installations under Section 800-10, which allows communication cable to be installed in the same manner as service drop electrical power conductor (Sec. 230-24) at lesser heights for sloped roofs.

9.7.4   Underground Communication Cable

According to NEC Article 800-11, underground circuits entering a building shall comply with 1 and 2 below.

1. Underground communication wires and cables in a raceway, manhole containing electric conductors, or nonpower-limited fire alarm circuit conductors shall be in a section separated from such conductors by means of brick, concrete, or tile partitions.
2. Where the entire street circuit is run underground and the communications circuit conductors within the block are placed as to be free from probability of coming in contact with power circuits over 300 V to ground, the insulation requirements of 800-12(a) and 800-12(c) shall not apply. Insulation supports shall not be required for the conductors, and bushings shall not be required where the conductors enter the building.

9.7.5 Communication Cable in a Conduit

Groups of cable that are run in a conduit should be pulled together. A pull line should always be pulled with a new cable to facilitate future cable installation within the conduit.

The cable manufacturer's guidelines for pulling, tension, and bend radius should be followed when a new cable is installed. The NEC guidelines should be followed for conduit size, and sufficient pull boxes should be installed in a conduit run to facilitate easy cable pulling. The conduit should probably be oversized to allow for future addition of communication equipment. The maximum total of all bends in a conduit run between pull boxes should be less than 360 degrees.

Whenever possible, twisted-pair and coaxial cable should run perpendicular to power cables. Electrical equipment such as circuit breakers, fluorescent fixtures, motors, printers, and transformers should be avoided to prevent electrical noise pick-up.

9.8 CABLE INSTALLATION TECHNIQUES

The remainder of this chapter is dedicated to examples of cable installation techniques and suggestions that might be used to facilitate cabling.

9.8.1 Above-the-Ceiling Installation

Figure 9-2 illustrates installation within a relatively small area above a ceiling.

The area below the ceiling in most buildings has a false ceiling for aesthetic appeal, improved acoustics, and air conditioning and heating efficiency. Light cables can be placed across the ceiling supports. Care must be taken not to overload the supports. Figure 9-2a illustrates a zone installation. The area is served by a cable from the closet to a multiple cable adapter from which drop cables are connected to the individual communication equipment. Figure 9-2b is an example of an over-the-ceiling home run in which each piece of communication equipment is connected directly to a source cable at the home closet or cabinet. Figure 9-2c represents the simplest method of adding a single piece of communication equipment. The cable is simply punched through the ceiling and connected to the communication equipment. No special fixtures, channels, or ties are used. However, whenever possible all cables should be supported by ties to structures other than the ceiling. The most permanent overhead installation and the most costly is to install conduit, as shown in Figure 9-3.

9.8.2 Under-Floor Runs

If buildings were designed with future needs in mind, there would probably be facilities for under-the-floor cabling. Figure 9-4 illustrates an under-the-floor conduit system. Hopefully, room was left for additional cables and a pull line was left in the conduit. If these two factors were met in the construction of the facility or past wire additions, the pulling of additional cables should be simple. All the cables should be pulled together. It is advisable to use an inert "cable grease" on the cables to prevent binding with other cables in the conduit. Cable grease looks like

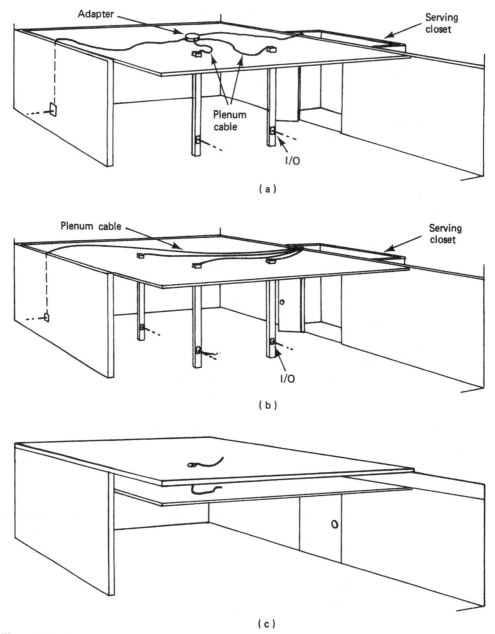

Figure 9-2 Installation within a small area above the ceiling: (a) a zone type of installation, (b) a home-run type of installation, (c) ceiling panel punch-through installation.

liquid soap, but it is a lubricant that will not harm the insulation on the cable and will dry in a short period of time.

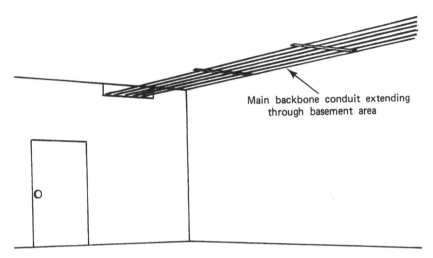

Figure 9-3 Conduit installation.

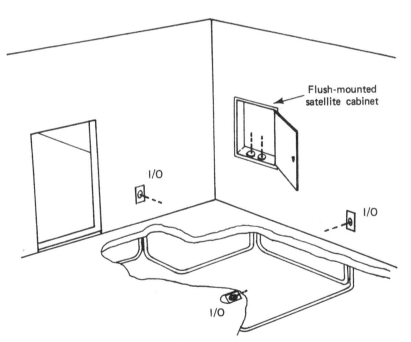

Figure 9-4 Under-the-floor conduit system.

Figure 9-5 depicts a wire mesh cable grip that can be used to pull cables through conduit. The grip is available in many sizes and must be the proper size for the cable or cables to be pulled. When multiple cables are to be pulled, the total diameter of the bundle must be measured and must fall within the circumference range of the cable grip.

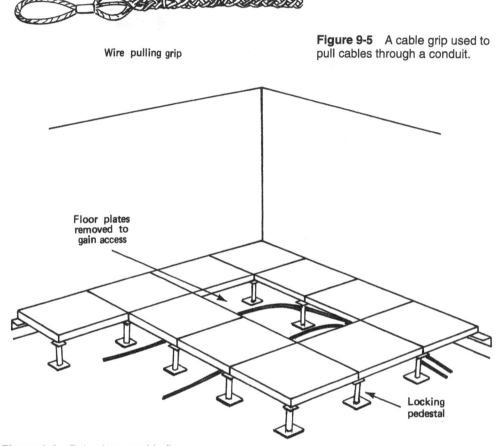

Wire pulling grip

Figure 9-5   A cable grip used to pull cables through a conduit.

Floor plates
removed to
gain access

Locking
pedestal

Figure 9-6   Raised removable floor.

Some buildings were designed with removable floor sections above the concrete floor (Figure 9-6). This type of floor can be added to almost any room and may be advisable where large numbers of wires are necessary to supply the communication system in the area.

Figure 9-7a shows an example of a cellular floor. Cells or tubes are placed in the floor support under the raised floor. Electrical and communication cables are run through the cells. Communication and electrical wires must not be run through the same cell. Figure 9-7b illustrates a double-suction-cup tool used to remove panels or cells of a raised floor.

Figure 9-8a illustrates under-the-floor ducting that was installed under a raised floor. Ducting was provided for both electrical and communication cables. Note that a separate closet is provided for the communication connections and an outlet box is provided for the electrical connections to keep the two isolated.

Figure 9-8b illustrates another example of under-the-floor ducting. This ducting is on one level and is necessary where there is insufficient space for two levels of ducting. The ducting has

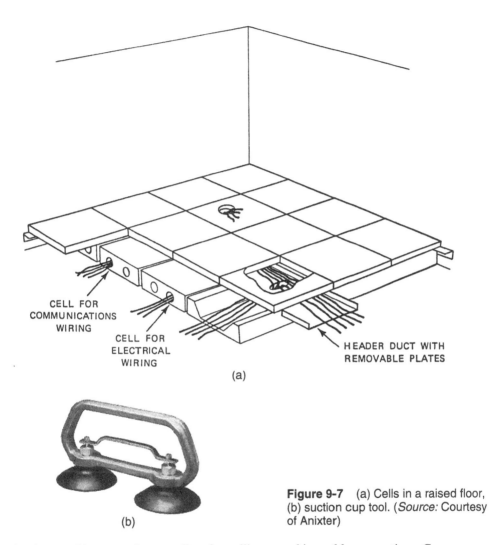

CELL FOR
COMMUNICATIONS
WIRING

CELL FOR
ELECTRICAL
WIRING

HEADER DUCT WITH
REMOVABLE PLATES

(a)

(b)

Figure 9-7   (a) Cells in a raised floor, (b) suction cup tool. (*Source:* Courtesy of Anixter)

junction boxes with access plates to allow for pulling or making cable connections. Copper communication cable and electrical cable should not be placed in the same duct.

9.8.3   Baseboard Installations

An example of baseboard installation is shown in Figure 9-9. Hollow baseboard can be installed in almost any area. The cable is routed behind the baseboard to the equipment. The cable can be run into another room by making a hole in the wall under the baseboard.

Figure 9-10 represents a variation of the baseboard method by utilizing hollow molding. The molding can be installed at any height on the wall or in the corner where the wall and ceiling meet.

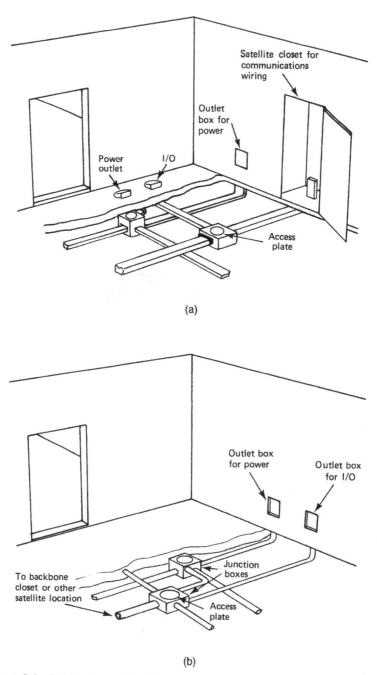

Figure 9-8 Ducting in a raised floor: (a) two-layer ducting, (b) single-layer ducting with junction boxes and access plates.

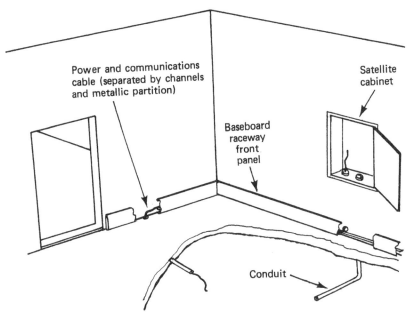

Figure 9-9   Baseboard installation.

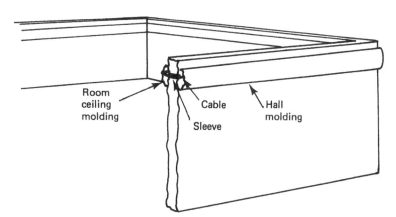

Figure 9-10   A molding cable installation.

Vertical connections can be made by using a wall strap, as shown in Figure 9-11. The cable is clamped to a steel support bracket that runs from the ceiling to the floor.

Whenever overhead cables are connected between buildings from a pole (Figure 9-12), a steel support strand must be included in the cable and secured to the pole and the building by appropriate hardware. If the cable is run down or along the building, it must be protected by a metal or plastic EMT cable guard. In some buildings, tunnels have been included for cable

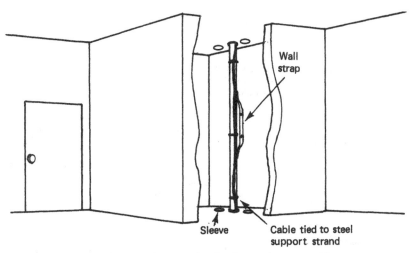

Figure 9-11   Vertical riser cable run that is clamped to a steel brace.

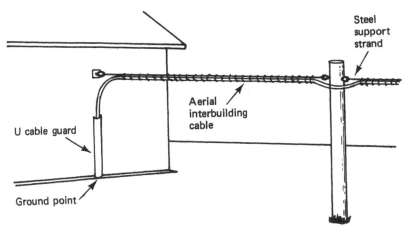

Figure 9-12   Cable drop from a pole to a building.

installation (Figure 9-13). Separation of electrical and communication cables is very important in a tunnel installation. NEC code must be followed and the cables must be magnetically separate to prevent electromagnetic pick-up.

Cabling can be purchased with an underground feeder (UF) insulation rating that can be directly buried in the ground (Figure 9-14). The UF-type insulation is a tough plastic that is not affected by water or acid in the soil. However, gophers and ground squirrels love it. The small additional cost of EMT-type conduit is a wise investment.

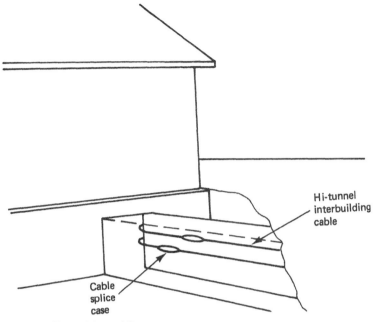

Figure 9-13 Tunnel for cabling.

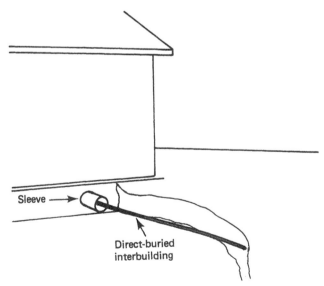

Figure 9-14 Special types of cables can be buried directly in the ground.

9.9 GENERAL RULES FOR CABLE INSTALLATION

In general, the installation techniques for installing twisted-pair, coax, and fiber cables are the same. Basic guidelines are as follows:

- Conduct a full and complete site survey prior to planning the installation.
- Develop a cabling pulling plan for each section of the cable installation.
- Follow pulling instructions for the particular type of cable: twisted pair, coax, or fiber.
- If conduit is to be used, do not exceed maximum recommended bends. If excessive bends are necessary, install pull boxes.
- Do not exceed the maximum bend radius of the type of cable being installed.
- Do not exceed the maximum recommended pull load for the cable.
- Document the installation.
- Test the installation.

NEC-approved cables must be approved for any indoor use. An approved firewall must be utilized for any penetration of a floor, wall, or ceiling. Cable fire code was covered in Chapter 5.

9.9.1 Site Survey

Survey of the site identifies the location of problems that require special care before cable installation. Physical hazards should be noted and avoided if possible. Copper cable and fiber cable with metallic strength members should always be kept away from power lines. NEC and other building code regulations must be considered and addressed. The fire marshal or local building department may have to be contacted. Hazardous materials, such as residential asbestos and fiberglass insulation, must be considered. Safety for installation and premise workers must be a prime consideration.

Every cable has a recommended minimum bend radius. During installation it is important that the cable never be bent or kinked into a radius less than the recommendation. During final placement the radius must be no more than the manufacturer's recommendation. Fiber as well as other cable can be coiled as long as the minimum radius guidelines are followed.

9.9.2 Cable Maximum Pulling Tension

The maximum pulling tension must not be exceeded for the cable at any time during installation. Most cables designed for outdoor use have a tension strength of at least 600 lb (272 kg). After the cable is installed the tension on the cable should be less than the maximum recommended tensile strength. Overhead outdoor cable should be clamped at intervals that will prevent unnecessary load on any part of the cable. Vertical installed cable must be suitably clamped at intervals that prevent both cable movement and cable weight from exceeding recommended long-term load.

Figure 9-15 Strength member tied directly to the pulling connector; cable end must be sealed for protection. (*Source:* Courtesy of Corning Cable Systems)

9.9.3 Pulling Instructions

The pulling instructions of the cable manufacturer should be followed to prevent distortion of the cable. Fiber cable demands special consideration to prevent elongation or breakage of the fiber. The strength member must be located and proper attachment must be made directly on the material to prevent stressing the fiber (Figure 9-15). For indirect attachment a pulling device such as the Chinese basket or Kellem's Grip (Figure 9-16) is connected over the entire cable.

The basket pulling device should spread the force over 1.5 to 3 ft along the cable. The basket device must be the proper size for the diameter of the cable to assure a positive grip. For assured grip and even pulling the basket should be taped to the cable. Care must be taken in pulling to prevent binds or kinks and excessive force. Friction between a cable and conduit can be reduced by use of lubricant that is compatible with the jacket of the cable. A cable pulley device preserves cable integrity, maintains the proper bend allowance, and allows one technician to feed the cable into the ceiling or around a bend.

As a rule it is best to install connections to cables after the cable is in place. This is particularly true for fiber as connection makes it more difficult to protect the fiber from stress. If the pull is to be entirely from one end, connectors can be safely installed on the other end. If the cable is factory manufactured with connectors on both ends, effort must be made to protect copper or fiber from stress and possible disconnection from the connector. Cushioned enclosures should be used to protect connector and cable during pulling. The lead end of the cable should be sealed to prevent the intrusion of water or other foreign materials.

Figure 9-16 A Kellem's Grip® connects to distribute the pulling force over the cable's outer surface. (*Source:* Courtesy of Corning Cable Systems)

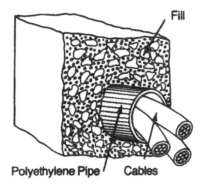

Fill

Polyethylene Pipe   Cables

Figure 9-17 PVC conduit used to protect underground cable. (*Source:* Courtesy of Corning Cable Systems)

Cable installations can be made bidirectional by laying the cable in figure-8 loops on a surface from which the cable can be pulled from both ends. A single cable should occupy less than 53% of the area of a duct or plenum, two cables less than 31%, and three or more cables less than 40%. When capacity remains in a conduit or plenum it is advisable to pull an extra cable or leave in a pull line for future needs.

9.9.4   Direct Burial

Cable designed for direct burial may experience water, pressure, crushing from rocks, ground displacement due to construction, and insulation removal by rodents. These hazards are usually prevented by burying the cable 3 to 4 ft. Direct plow-in requires cables that can withstand irregular pulling forces. Double jacking, gel lining, and metal sheathing are some of the schemes used to prevent damage to direct burial cable. It is advisable to use conduit when burying cable. Polyethylene PVC conduit (Figure 9-17) is an inexpensive protection that allows the use of less expensive cable types than direct burial types.

9.9.5   Aerial Installation

Cables for long outdoor overhead installation are usually temperature stabilized by the use of a stabilizing unit. The stabilizing unit may be of steel or fiberglass/epoxy, depending on the potential for lightning or electrical hazards. Figure 9-18 depicts an aerial connection for an overhead fiber-optic cable.

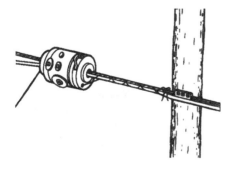

Figure 9-18 Aerial connection for a fiber-optic cable. (*Source:* Courtesy of Corning Cable Systems)

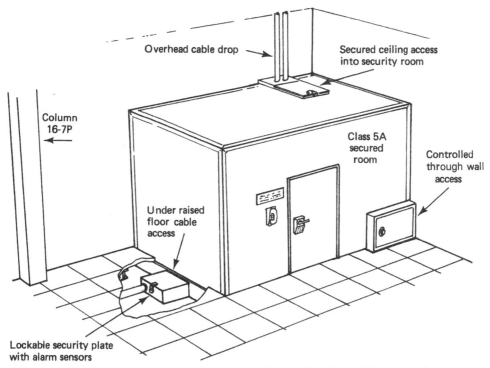

Figure 9-19 Designated controlled access room. (*Source:* Courtesy of Anixter Inc.)

9.10 CABLE SECURITY

Under normal conditions, typical and usual methods of cable installation, access, and routing are applicable. However, in the case of access into a controlled environment, a more restrictive set of access rules apply. For example, a research and development group laboratory or competitive-analysis room within a controlled space would generally be considered as a *controlled area.* Corporate and site guidelines must be developed and implemented to restrict access to cabling to these functions by controlling access to cable trays and overhead cable support systems. Approved methods and access points must be established prior to cable routing and installation. Figure 9-19 depicts an example of a controlled access room. The security system to the area can vary from a simple lockable cable feeder opening to dedicated individual fiber and cable lines, each with an alarm-equipped access cover and video surveillance in place.

The three areas of exposure that need to be addressed when dealing with security or restricted access rooms are as follows:

1. Below secured room access (cable trays under a raised floor)
2. At secured room wall access
3. Above secured room access (through overhead cable and wire trays)

Cable patch panels and breakout boxes may also need to have some type of security system.

9.11 CABLE INSTALLATION HARDWARE

There are many manufacturers of cabling hardware and thousands of different hardware pieces each designed to make the job of a cable installer easier. Before attempting a cabling installation or addition or modification of a cabling system, the installer should review several catalogs and become knowledgeable about cabling hardware. Most manufacturers or distributors will furnish, on request, details on the installation of their products. In this chapter we will review only a sampling of these products as examples of cabling hardware. A number of hardware manufacturing firms is listed in the Appendix at the end of the text.

9.11.1 Overhead Cabling Hardware

Most office buildings have a false ceiling installed several feet below the roof for aesthetics, acoustics, soundproofing, heating, and cooling. The "dead space" above the ceiling is used to conceal the heating and cooling ducts, electrical conduit, water lines, and gas lines. This space is an ideal location for most communication cables.

When an area requires only a few cables, they can probably be placed directly on the ceiling supports. However, as the telecommunication system grows and the cable requirement increases, more support is necessary. Figure 9-20 illustrates a cable runway that is supported by overhead brackets. Brackets can be added to the cable runway that enable change of elevation or direction as desired.

When an area requires a large number of communication and data line drops, a grid support structure is more suitable. The grid structure depicted in Figure 9-21 has a hatch installed for overhead access. Both the cable runway and the grid structures can be supported by threaded rods from the roof or by floor column supports.

9.11.2 Central Distribution Points

As a telecommunication network grows, there is increased need for a large number of user terminals and other equipment to access the host computer. A growing network requires a central information access point for cabling, whether it be twisted pair, coaxial, or fiber optic.

The most common central distribution point is the terminal block used for twisted pair simply because it has been used for many years by telephone companies. Figure 9-22a depicts a 66-terminal connectorized punch-down block. The numbering convention for such a block is illustrated in Figure 9-22b. In a small facility a block may be fastened to a wooden wall by screws. However, larger facilities such as depicted in Figure 9-23 require that methods offering easier cable identification and better tracking control be utilized.

Figure 9-24 illustrates a typical communication rack that is loaded with relay racks. The distribution racks should be installed in a preselected dedicated area (usually called a closet). The closet should be clean, dry, and cool, with ample space for cable installation, modification,

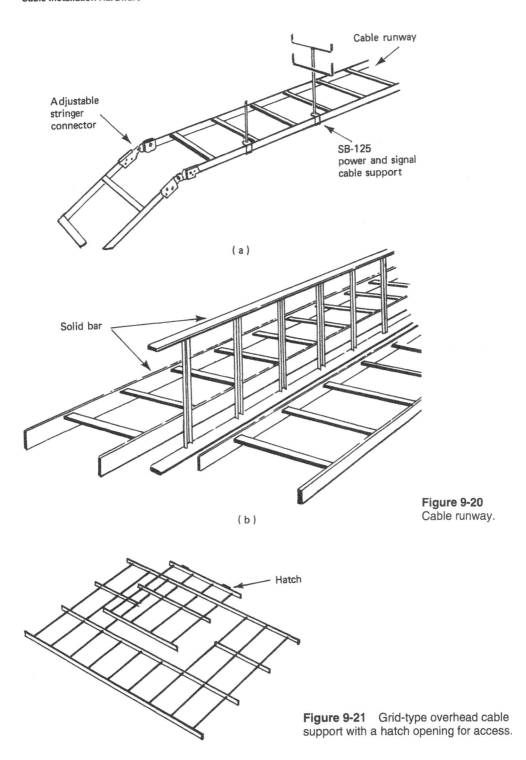

Cable runway

Adjustable
stringer
connector

SB-125
power and signal
cable support

(a)

Solid bar

Figure 9-20
Cable runway.

(b)

Hatch

Figure 9-21   Grid-type overhead cable
support with a hatch opening for access.

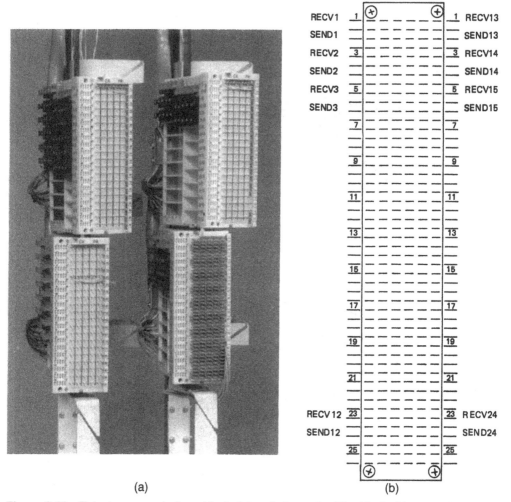

Figure 9-22 Telephone punch-down block: (a) a photograph of the block, (b) a sample numbering system for the block.

and testing. Space should also be dedicated for future expansion of the network. A small area may "make do" with a rack cabinet.

Terminal blocks with punch-down connections are common in older facilities. However, designers of newer installations are looking to different and better methods for cable connectability. The two connectors utilized most often for twisted-pair cabling are the RJ11 (6 pin) and the RJ45 (8 pin) connector. An RJ11 extension cord is shown in Figure 9-25. These connectors have become standards in modular telephone systems and are often used as the input and output for PCs and other digital equipment. Their adoption was spurred when manufacturers realized

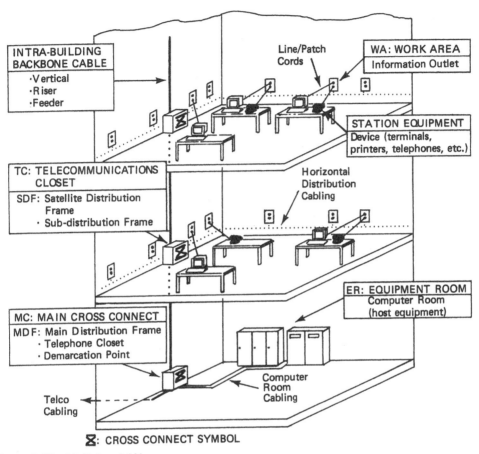

Figure 9-23 Multistory LAN.

that only a few of the 25 pins were being used in the EIA RS232 DB25 connectors (Figure 9-26).

Patch panels utilizing the RJ telephone connectors allow easy cable rerouting without the need for tools.

Coaxial cables are terminated with either BNC or TNC connectors and have different characteristic impedances than twisted-pair or twinaxial cable. To prevent a mismatch of imped-ance and a corresponding signal attenuation, an impedance-matching device called a balun (Fig-ure 9-27) is connected between the TWP and the coaxial cable. A connectorized balun can be connected at any point between twisted-pair and coaxial cable. The baluns can be mounted at a cable panel that can be mounted in a rack with punch-down blocks, between the workstations and the wall plates, and so on. Figure 9-28 illustrates a system utilizing twisted-pair and coaxial cable with a balun-matching connector panel.

Figure 9-24 Communication cable rack with terminal blocks installed.

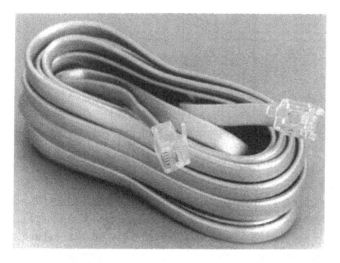

Figure 9-25 RJ11 extension cord.

Coaxial cables are rather heavy and should never be hung on their terminal connections or stretched. Strain relief should always be provided to prevent the cable from separating from the connector. Cable ties should be used to fasten coax to a solid support both in the ceiling and

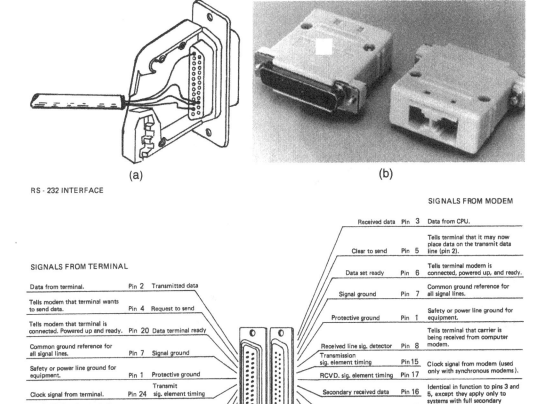

Figure 9-26 EIA RS232 DB25 connector: (a) most applications utilize only a few of the connector pins, (b) appearance of connector, (c) EIA RS232 pin assignments.

especially when the cable is run up walls between floors. The ties should be tight enough to hold the cable but not so tight as to crush the cable. If the tie is too tight, the dielectric insulation will be crushed and the outer shield may short to the inner conductor. When coaxial cables are connected into a patch panel a strain loop must be provided. Figure 9-29 illustrates a patch panel cable guide and the correct method of providing strain relief.

Figure 9-27 Balun impedance-matching device.

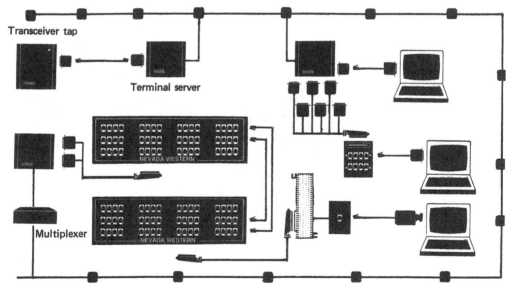

Figure 9-28 System utilizing twisted-pair and coaxial cable. (*Source:* Courtesy of Nevada Western Corp.)

Care must be taken when coax is laid on a ceiling support or an overhead cable grid so as not to strain the cable as it is being pulled across the overhead supports. This can be prevented by using cable pulling loops, as depicted in Figure 9-30.

Fiber-optic cabling requires fewer actual cable home runs than either twisted-pair or coaxial cable. However, the electrical signals at the source must be converted to light signals for the fiber cable and reconverted to electrical signals at the destination. This can be accomplished

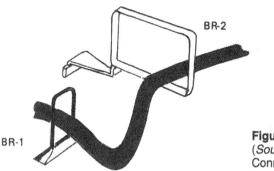

BR-2

BR-1

Figure 9-29 A coax patch panel cable guide. (*Source:* Courtesy of Automatic Tool and Connector Company)

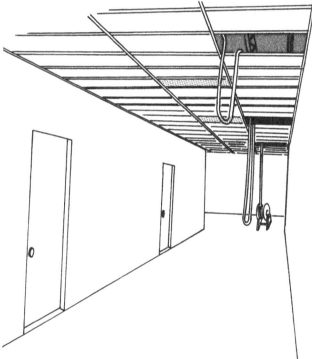

Figure 9-30 Cable pulling loops will prevent strain on the cable and cable connectors.

by a mux at each data processing device or at a central distribution panel. The latter is by far the most economical. Figure 9-31 illustrates a multistory-multibuilding LAN that utilizes all three communication wiring media. A single fiber-optic cable connects the buildings, and coaxial and twisted-pair cables connect the telephone closet and various racks on each floor.

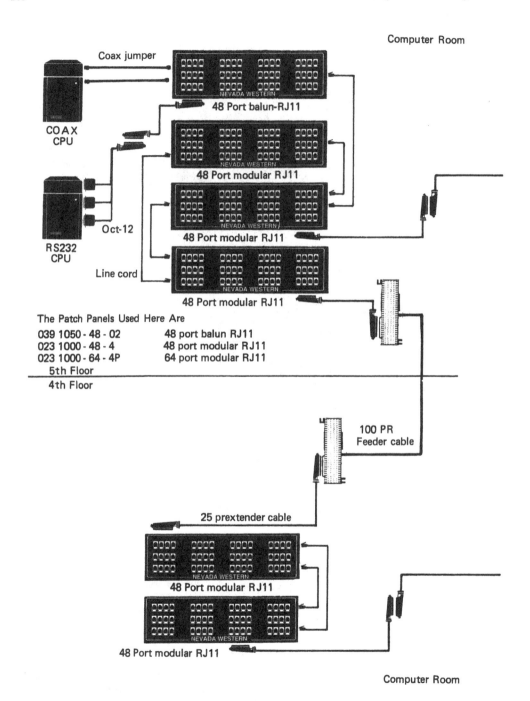

Figure 9-31 Multistory-multibuilding wiring system for a LAN.

Telephone Closet

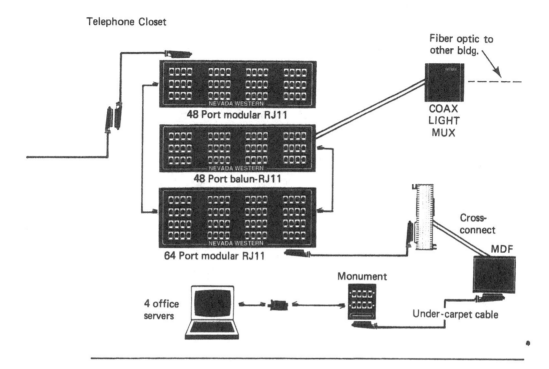

4 office servers

Monument

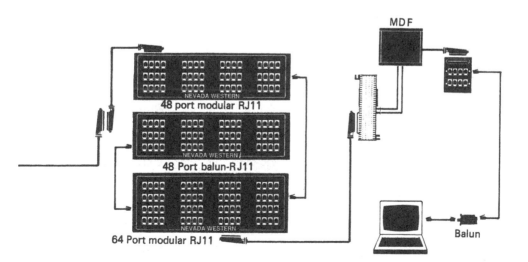

Telephone Closet

Figure 9-31 Multistory-multibuilding wiring system for a LAN. (Continued)

9.12 GROUNDING THE CABLING SYSTEM

We have mentioned the necessity for proper grounding of equipment and cabling several times in this text; however, it is such an important topic that it bears repeating here. The grounding of equipment is primarily for the safety of the workers. An electrical short in ungrounded equipment can result in the outer case being at line voltage with respect to another piece of grounded equipment, a conduit, the earth, or a concrete floor.

As stated earlier, electrical noise is the introduction of any interfering voltage developed internally or externally to an equipment or transmission line. The larger the number of electrical/electronic devices within a premise, the greater the possibility for electrical noise within the system. The most likely pick-up of electrical noise would be as follows:

1. Low-frequency interference (LFI), usually 60 H electromagnetic radiation from power lines, fluorescent lights, and so on
2. Electrostatic discharge interference (EDI) from electrostatic voltages caused by the motion of people or rotating electrical devices
3. Electromagnetic interference (EMI) from electronic devices such as computers, or other data transmission lines
4. Radiated frequency interference (RFI) from signals in the air such as radio signals or microwave signals

Grounding of the plant, the equipment, and the cabling must meet the NEC requirements for proper grounding. The latest edition of the NEC manual should always be consulted because the requirements are sometimes changed or modified.

Review Chapters 1 through 4 for the proper grounding techniques for twisted-pair, coax, and fiber-optic cabling.

9.13 ELECTRICAL PROTECTION DEVICES

Electrical and overvoltage protection is divided into two categories, primary and secondary protection. Primary protection devices are designed to protect people and buildings and are usually installed by the local exchange carrier on the regulated side of the network. Primary protection activates when lightning strikes, power lines cross, or other events cause high voltages to occur. The protection devices are triggered to divert the high voltage and current to ground. Primary protection devices do not respond fast enough or at a level that will protect most electronic equipment.

Secondary protection devices are installed behind primary protection and stop any damaging surge voltages or currents that pass the primary protection. Secondary protection devices are installed between the building entrance and the system equipment and as close as possible to the equipment. Figure 9-32 illustrates two applications of secondary voltage and current protection. Figure 9-32a illustrates the correction location for surge protection between the building entrance and the system. Figure 9-32b illustrates the correct location for surge protection

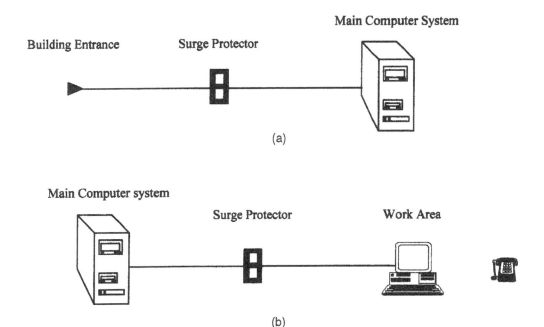

Figure 9-32 Secondary protection: (a) protection between the building entrance and the system, (b) protection between the system and work area.

between the system and work area equipment. In either case the device should be as close to the equipment being protected as possible.

Primary protection devices are a form of lightning arrestor that may utilize either semi-conductor or gas tube overload devices to short the high voltage and current of a lightning strike to ground. The breakdown voltage and reaction time of the device is of primary importance to prevent circuit damage. The voltage level of the protection device is found by measuring the DC and AC voltage levels at the equipment to be protected.

The effective (RMS) value of the AC voltage is multiplied by 1.414 to obtain the peak value. These two voltages are added together to obtain the maximum possible voltage at the equipment. For example, suppose the DC voltage measured 24 V and the AC voltage measured 120 V. The maximum voltage would be

$$V_{max} = V_{dc} + V_{ac\ peak} = 24 + (120 \times 1.414) \cong 169.7V$$

Secondary protectors are comprised of semiconductor devices that clamp voltage surges and leakage current to maximum safe values and prevent damage to electronic switches, digital and hybrid ports, and terminal electronics. Figure 9-32 depicted examples of surge protectors used to protect a main computer and a work station. Figure 9-33 depicts a multitasking network with AC surge protection, data surge protection, and ground monitor. The protection devices are located at various points in the network to afford maximum protection to the system.

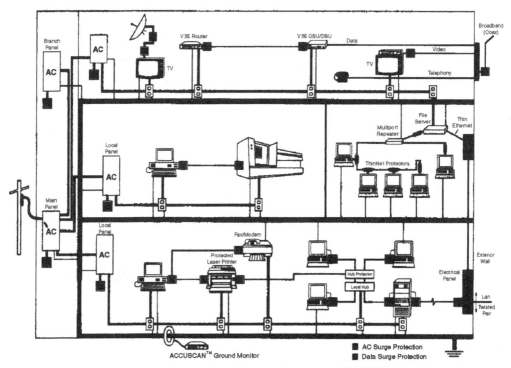

Figure 9-33   Surge and ground protection. (*Source:* Courtesy of TYCO Electronics Inc.)

As mentioned several times previously, grounding is of utmost importance for both primary and secondary safety. A protection device is only as good as its ground. This means that every protection device must be bonded to an approved building ground.

9.14 SUMMARY

There are as many cable schemes as there are buildings. The building topology, the present inventory of data processing equipment, the current in-place wiring, and the future telecommunication plans will all dictate the type of network and wiring media employed. The wiring media and wiring methods should be tailored to the facility and the present and future needs of the business. The wiring of a network should be given as much attention as is given to selecting the data processing equipment.

QUESTIONS

1. Why should all cabling be pulled in a conduit at one time?   ·
2. From the point of view of the wirer, compare the installation problems of twisted-pair, coax, and fiber-optic cabling.

3. How would you determine which cables are out of use in your facility?

4. Compare the procedures of finding an unused cable and adding a new cable for a user.

5. Why is the proper termination of a coax cable important?

6. What are the proper methods for pulling fiber-optic cables?

Premise Wiring Systems

10.1 INTRODUCTION

There are many ways to design premise wiring for any LAN or WAN. The final plan will, of course, reflect the legacy of equipment, systems, products, and wiring of the facility. However, the designer is an artist that manipulates the old and the new with current and future needs to develop the final design. The final architecture will also reflect to some extent the cabling, connectors, and devices of the hardware manufacturer.

The most important decision the telecommunication manager must make involves selecting the structured cabling system. The medium should support data, voice, video, and telemetry/ sensor applications, for current or future planning. Proper connectivity that allows for manageability, flexibility, versatility, and future expansion should be designed into the system. To be universal to all possible applications the cabling plan must support all logical configurations, such as point to point, bus, star, tree, ring, and hybrid.

Before examining network designs, we must become familiar with cabling terminology:

- **Backbone cable** is the portion of premise cabling that provides connections among telecommunications closets, equipment rooms, and entrance facilities. Backbone cabling consists of transmission media, main and intermediate cross connects, and terminations for the horizontal cross connect, equipment rooms, and entrance facilities. Backbone cabling can also be classified as campus backbone (cabling between buildings) or building backbone (cabling between floors or closets within a building).
- **Intermediate cross connect (IC)** is a secondary cross connect in the backbone cabling used to terminate and administer backbone cabling between the main cross connect and the horizontal cross connect.

• **Horizontal cross connect (HC)** is a cross connect of horizontal cabling to other cabling (for example, to horizontal backbone or equipment). Single-user cable can be installed for permanent office installations. The number of fibers or other cables depend on the number of applications needed by the user. Multiuser open office application requires high cable installation to allow for rearrangement or expansion of applications.

• **Campus backbone cable** is cabling between buildings and requires special consideration for current and future usage because of the expense in installing interbuilding pathways and disruption of traffic during trenching and tunneling. Table 10-1 gives the solution for a campus backbone. It is better to err on the side of excess since the cost of the fiber is the least of installation expenses. The example in Table 10-2 employs extensive applications for which the designer utilized a 60-fiber hybrid cable comprised of 48 multimode fibers and 16 single-mode fibers. The designer has allowed 16 spares for future expansion.

Table 10-1   Fiber count for a campus backbone example

Application	Number of MM Fibers	Number of SM Fibers	Comments
ATM	2	2	62.5 μm up to 155 Mbps SM for < 622 Mbps
Video (broadcast)	2	2	62.5 μm up to 4 channels SM for 24 channels
Video (interactive)	2	0	—
Fibre Channel	6	0	3 channels
100BaseF (Ethernet)	2	0	—
Token ring	4	0	—
FDDI	4	0	—
Voice	4	0	DS3 rate
Security (interactive)	4	0	—
MM spares (100% rule)	16	0	—
SM spares (100% rule)		4	—
Total	48	12	

MM = Multimode, SM = Single mode.

Source: Courtesy of Siecor Building, Inc.

In this chapter we will present several manufacturers' conceptions of LANs using their hardware and technology. All are of excellent design and offer many useful ideas for the tele-communication professional.

Table 10-2   Building backbone design

Application	Number of MM Fibers	Number of SM Fibers	Comments
ATM	2	0	62.5 μm up to 622 Mbps
Security video	2	0	Remote control—2F
1000BaseF (Ethernet)	2	0	—
Token ring	4	0	—
Voice (remote PBX)	4	0	—
FDDI	4	0	—
MM spares (50% rule)	9		

Source: Courtesy of Siecor Building, Inc.

10.2 FIBER-OPTIC NETWORKS

The primary planning of a wiring system[*] involves two basic parts, the physical topology and the logical topology. The physical refers to the arrangement and type of cabling to be employed. Figure 10-1 illustrates these concepts. Figure 10-1a is the logical aspects of the flow within the system; Figure 10-1b details the physical cabling layout.

One common network topology is point to point (page 135, Chapter 6), where two nodes require direct communication. An example is depicted in Figure 10-2, where two fibers are used to carry signals between two applications, one to transmit and one to receive. Common applications of point to point are terminal multiplexing, telemetry/sensor, Fibre Channel, and satellite up/down links.

The logical star topology is an extension of point-to-point topology, where a collection of links has a common node that controls the system. Star applications include PBX, ATM, data switch, a video security system with a central monitoring station, and a multilocation interactive video conference system. Figure 10-3 depicts an example of logical star topology.

Logical ring topology (Figure 10-4), supported by token ring (IEEE 802.5) and FDDI (ANSI X3T), is a very common data communications topology. In ring topology nodes are con-

[*] Siecor Design Guide Release 4, Siecor Operations, LLC, P.O. Box 489, Hickory, NC 28603-0489 USA
  http://www.siecor.com 1-800-Siecor5.

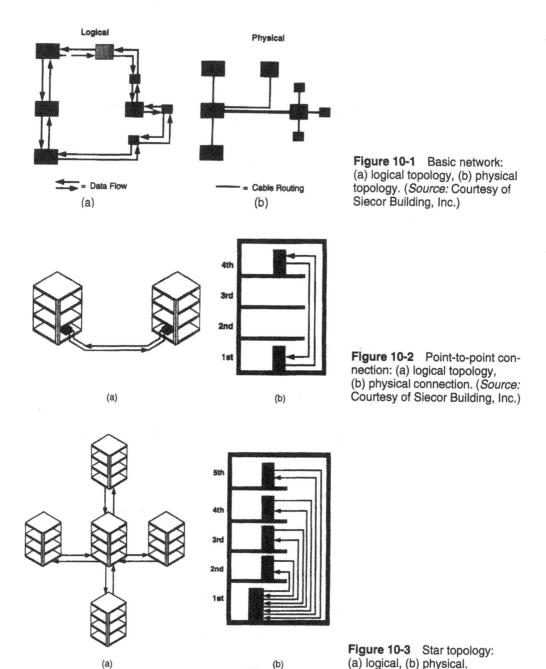

Figure 10-1 Basic network: (a) logical topology, (b) physical topology. (*Source:* Courtesy of Siecor Building, Inc.)

Figure 10-2 Point-to-point connection: (a) logical topology, (b) physical connection. (*Source:* Courtesy of Siecor Building, Inc.)

Figure 10-3 Star topology: (a) logical, (b) physical.

nected to an adjacent node in single or dual counterrotating rings rotating in opposite directions. The advantage of this topology is that if one node fails, one ring will automatically switch to the other, allowing normal operation of the remaining nodes. This enhancement requires two pairs

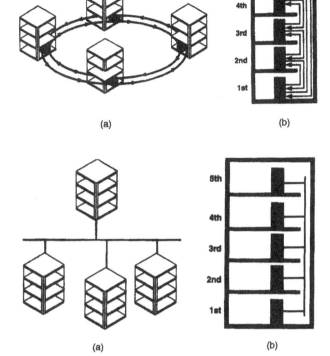

(a)   (b)

Figure 10-4 Counterrotating ring topology: (a) logical topology, (b) physical topology. (*Source:* Courtesy of Corning Cable Systems)

Figure 10-5 Bus topology: (a) logical diagram, (b) physical diagram. (*Source:* Courtesy of Corning Cable Systems)

per node rather than a single pair as in a simple ring. Typically, FDDI employs a counterrotating ring for the backbone and a single ring for the horizontal.

Bus topology is used for data communications and is supported by IEEE 802.3 standards. Figure 10-5 depicts bus topology in which all nodes share a common line. Transmission occurs in both directions on the line as opposed to a single direction on ring topology. When a node transmits, all nodes receive the information almost instantaneously. Applications employing bus technology are token bus, Ethernet, and Manufacturing Automation Protocol (MAP).

When bus or ring topology is implemented with a concentrator or an intelligent hub the device establishes logical electronic solutions designed to interface with a star network. This device places the bus or ring network in its backplane, resulting in connections being made at single or multiple locations. The physical connection of these networks appears to be a star technology and is best supported by a star cabling system. Table 10-3 compares the characteristics and notes the advantages of star and ring technology implemented by fiber media.

Table 10-3   Comparison of star and ring technology

Physical Star	Physical Ring
+ Flexible, supports all applications and topologies	– Less flexible
+ Acceptable connector loss with topology compatibility	– Unacceptable connector loss if star or bus logical topologies are required
+ Centralized fiber cross connect makes administration and rearrangement easy	– Connectors located throughout the system, making administration and rearrangement difficult
+ Existing outside ducts are often configured for physical star, affording easy implementation	– Physical implementation can be difficult, often requiring new construction
+ Easily facilitates insertion of new stations or buildings into the network	– Insertions of new stations or buildings into a ring can cause inefficient cable use or disputation to current applications
+ Supported by TIA/EIA-568A Commercial Building Telecommunications Cabling Standard	– Even with ring topology applications, most, if not all, nodes must be present and active
+ Cut cable will result in node failure without redundant routing	+ Node survivability in event of cable cut if cable utilizes redundancy
+ More fiber length is required than with logical ring topology	+ Less fiber length between nodes

Source: Courtesy of Siecor Building, Inc.

10.3 *TIA/EIA-568 COMMERCIAL BUILDING TELECOMMUNICATION CABLING STANDARDS*

The TIA/EIA-568 solution, depicted in Figure 10-6, is based on a hierarchical star for the backbone and a single star for horizontal distribution. The rules for backbone cabling include a 2000-m 62.5/125 µm multimode (3000-m for single mode) maximum distance between the main cross connect (MC) and the HC and a maximum of one IC between the HC and the MC. The MC is allowed to provide connectivity to any number of ICs or HCs, and the ICs are allowed to provide connectivity to any number of HCs. The standard does not distinguish between inside (facility) or outside (campus) backbones since these are determined by campus or facility size and layout. In most applications the MC to IC is the campus backbone link and the IC to HC the building backbone link, as in Figure 10-7. The exception would be a multistory building in which the backbone may be entirely within the building, eliminating the need for an IC.

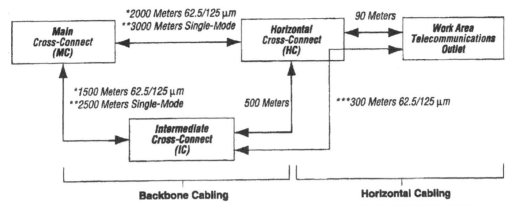

Figure 10-6   Commercial building telecommunications cabling standards (TIA/EIA-568A). (*Source:* Courtesy of Corning Cable Systems)

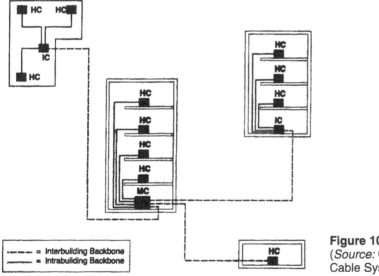

Figure 10-7   Backbone cabling. (*Source:* Courtesy of Corning Cable Systems)

The horizontal cabling (Figure 10-8) is to be a single star linking the horizontal cross-connect closet work area outlet with a distance limitation of 90 m. Distance limitation is based on copper data rate limitations. TIA published TSB-75 and TSB-72, which allow better utilization of optical fiber's high-performance characteristics in horizontal, multiuser outlet, and centralized cabling.

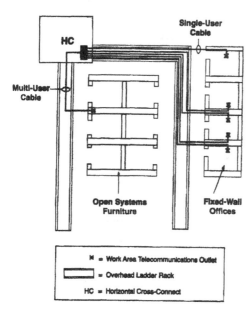

Figure 10-8   Horizontal cabling.
(*Source:* Courtesy of Corning
Cable Systems)

10.4 CAMPUS BACKBONE

Campus backbone cabling allows most options for versatility of any part of the network, espe-cially in large networks such as a college, industrial park, hospital complex, or military base. However, the campus backbone is also constrained by physical barriers such as right-of-way and conduit and duct availability.

In smaller networks with less area or fewer buildings the best plan involves linking all buildings requiring optical fiber to the MC. The cross connect in each building then becomes the IC that links the horizontal cross connect of each building, as depicted in Figure 10-9. The MC should be in close proximity or in the main equipment room. The MC, if possible, should be centrally located among the buildings served and have space for cross-connect hardware and equipment and suitable pathways linking other locations. The design in Figure 10-10 complies with the TIA/EIA-568A standard. The hierarchical star as the campus backbone has the follow-ing advantages:

1. Provides a single point of control for network management
2. Allows for testing and reconfiguration of system topology from the MC
3. Allows easy maintenance against unauthorized access for security
4. Allows easy addition of future campus backbone links

A design for large campus networks (in number of buildings or area) may require a two-level hierarchical star as depicted in Figure 10-10. The design utilizes selected ICs to serve a number of buildings. The ICs are linked to the MC. This option is effective when all cable can-not be linked to an MC or when it is desirable to segment the network by function, allowing

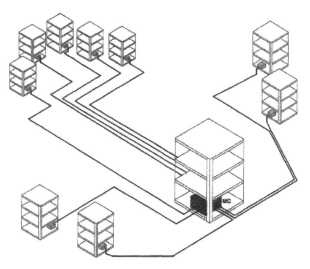

Figure 10-9 One-level hierarchy campus backbone. (*Source:* Courtesy of Corning Cable Systems)

more effective use of multiplexers, routers or switches, or other electronics for better utilization of bandwidth or cabling. It is recommended that when a two-level hierarchical star is used for a campus backbone, it is implemented by a physical star in all segments for flexibility, versatility, and manageability. There are, however, two conditions when a physical ring may be considered for linking interbuilding ICs and the MC: when the existing conduit supports it and when the network topology is token ring or FDDI. Siecor network designers, for example, seldom recommend connecting outlying buildings in a physical ring. Siecor recommends that "when a hierarchical star is used for a campus backbone, . . . it be implemented by a physical star in all segments." Figure 10-11 depicts a campus backbone showing recommended and nonrecommended connections.

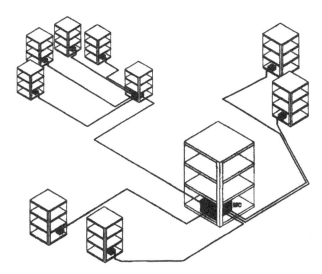

Figure 10-10 Two-level hierarchy campus backbone. (*Source:* Courtesy of Corning Cable Systems)

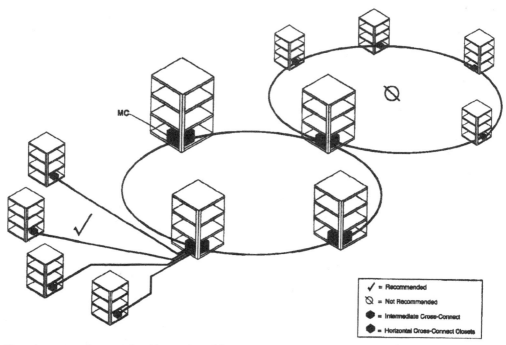

MC

✓

✓	= Recommended
Ø	= Not Recommended
●	= Intermediate Cross-Connect
⬟	= Horizontal Cross-Connect Closets

Figure 10-11   Campus backbone ring. (*Source:* Courtesy of Corning Cable Systems)

10.5 EQUIPMENT ROOMS AND NETWORK CABLING

The main cross connect ideally is located in the equipment room with the data center, security monitoring equipment, PBX, and other active equipment. Physically this may not be possible when multiple equipment rooms are served. While it is not recommended that there be more than one MC, connection to equipment rooms can be provided with fiber in separate sheaths or combined with the backbone fiber, as depicted in Figure 10-12.

10.6 SPLICE POINTS

The physical design of the cabling system should minimize splices. Splicing is time consuming and the source of possible attenuation losses. The small size and low cost of fiber is outweighed by the cost of splicing and possible faults. Figure 10-13 depicts examples of recommended and nonrecommended applications of cable connection.

Planning the location of splice points can result in considerable savings of labor cost because it is less expensive to perform all splices at one location than at several. This is true even when additional fiber is required. This concept is depicted in Figure 10-14, where additional cables are run to termination through splice points.

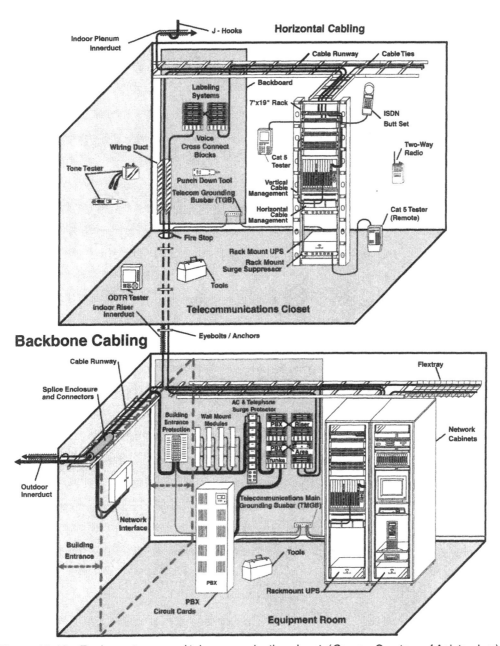

Figure 10-12   Equipment room and telecommunication closet. (*Source:* Courtesy of Anixter, Inc.)

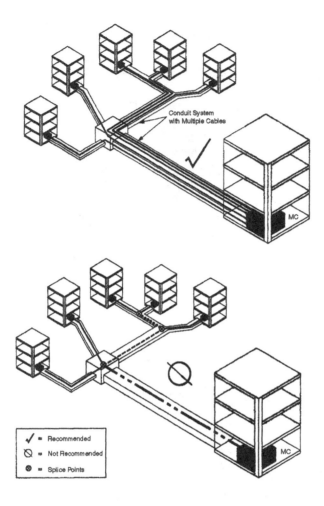

Conduit System
with Multiple Cables

MC

✓ = Recommended
⊘ = Not Recommended
⊛ = Splice Points

MC

Figure 10-13 Multiple cable lays. (*Source:* Courtesy of Corning Cable Systems)

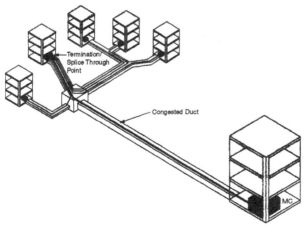

Termination/
Splice Through
Point

Congested Duct

MC

Figure 10-14 Consolidated splice point. (*Source:* Courtesy of Corning Cable Systems)

10.7 BUILDING CABLING ARCHITECTURE

As a summary to illustrate accepted wiring architecture we will present two building cabling schemes[†] for a multistory, multiuser building as envisioned by a global network design organization and telecommunication hardware distributor. The two systems are representative of skillfully designed architecture. The premise cabling solution will differ for each company, depending on the nature of the business, the equipment in place, and the existing wiring. The final cabling architecture should be flexible for future equipment addition, be manageable for the technical staff, meet compliance standards, and offer low overall system cost.

The following are two wiring solutions to a multistory facility: **centralized network administration** (CNA) and **distributed network administration** (DNA). The cabling system includes horizontal distribution methods to the work area, *single-user cabling* (SUC) and *zone distribution cabling* (ZDC). Interbuilding and interbuilding backbone cabling are also employed.

The difference between CNA and DNA is the arrangement of the LAN electronics and cabling connectivity used in support of the LAN. The CNA cabling system has all the LAN electronics in a centralized location. On the other hand, the DNA has the LAN electronics distributed throughout the building.

Unshielded twisted-pair (UTP) wiring or optical fiber may be used in a CNA arrangement in a small facility where all the users are within 90 m of the centralized closet. For larger premises, where connection distance exceeds 90 m, optical fiber must be used for data and video and UTP may be used for voice. The cabling distance requirements for DNA connection are the same as for CNA. However, all users must be within 90 m of the telecommunication closet.

10.8 CENTRALIZED NETWORK ADMINISTRATION

CNA places all connectivity in one centralized equipment room (closet). This facilitates the support of interdepartmental LAN work groups, reduces the requirement for closet space, increases LAN utilization, and forms a migration path for the utilization of higher-speed networks.

CNA is best for organizations that are concerned with long-term cost rather than initial installment, that desire central control of the LAN and flexible migration to high-speed networks, that migrate to or employ switched versus shared LAN, and that have inadequate closet space for housing LAN electronics. Figure 10-15 depicts a simplified plan for a CNA application. The architecture shown has been recognized by TIA/EIA with publication of TIA/EIA TSB72, Centralized Optical Fiber Cabling Guidelines.

10.9 DISTRIBUTED NETWORK ADMINISTRATION

DNA is an architecture that best addresses the needs of organizations that do not want their LAN centralized or whose LANs are under departmental control. The architecture also provides

---

† Premises Cabling Solutions, AMP Corp., P.O. Box 55, Winston-Salem, NC 27102-0055.

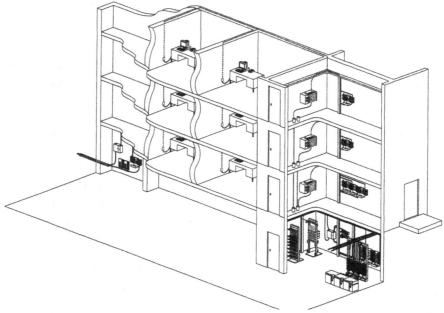

Figure 10-15   Centralized network architecture. (*Source:* Courtesy of TYCO Electronics Corp.)

- Better support for multitenant buildings without shared facilities where each tenant has LAN control
- The lowest installation cost
- Support for department-controlled LANs rather than corporate-controlled LANs
- Support of existing copper-based LAN wiring

Figure 10-16 depicts an example of a DNA architecture for the same facility that was shown in Figure 10-15. Notice that the entrance facility is the same for both types of architecture. An expanded view of the entrance facility is shown in Figure 10-17. The inserts, on the drawing, illustrate examples of cabling, cabling connectors, and closet hardware.

The main cross-connect facility is also shown as the same for both architecture schemes. This is the heart of both systems. The insert drawings depict typical connectors and splices, cable configuration, patch cables, and cross-connecting hardware.

Figure 10-18 illustrates an example of a telecommunication closet for a single department or a common group of users for the CNA system. The leftmost inserts depict typical workstation outlets and typical wiring media. The rightmost inserts depict typical closet connections necessary to connect the data equipment to the main cross connect and the hardware necessary to connect cable.

Figure 10-19 is an example of a single-user telecommunication closet using the DNA architecture. The equipment necessary for a single department or tenant of the DNA architecture

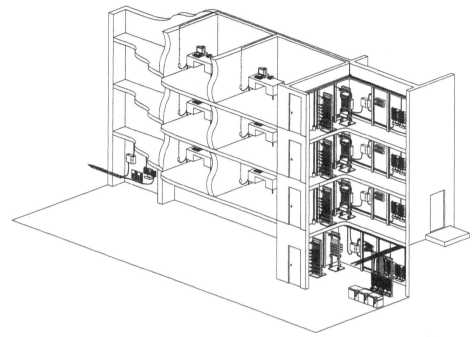

Figure 10-16 Multistory distributed network administration. (*Source:* Courtesy of TYCO Electronics Corp.)

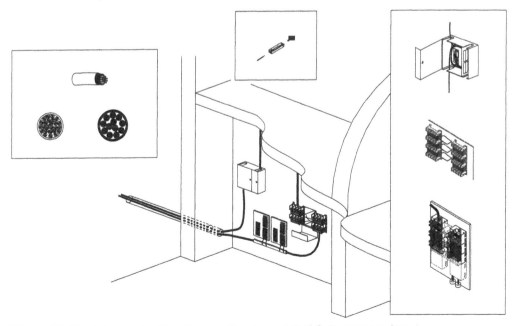

Figure 10-17 Entrance facility. (*Source:* Courtesy of TYCO Electronics Corp.)

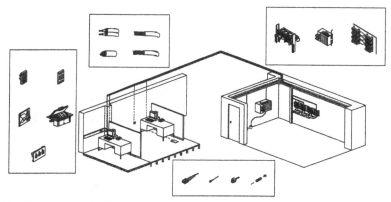

Figure 10-18 Telecommunication closet for a single unit. (*Source:* Courtesy of TYCO Electronics Corp.)

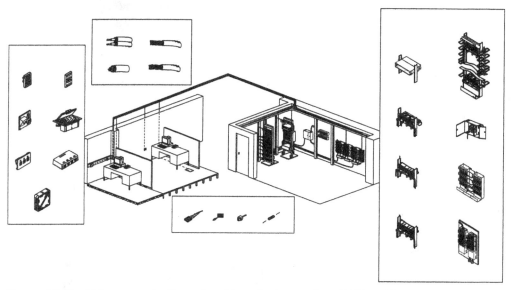

Figure 10-19 Telecommunications closet to a single user for a DNA architecture. (*Source:* Courtesy of TYCO Electronics Corp.)

system is much more complex than that of the CNA architecture because much of the main cross-connect functions is transferred to each entity. Here again, examples of cabling, connections, and hardware are depicted in the inserts.

An example of zone distribution cabling necessary for a floor of a multistory building, a department, a laboratory, or a tenant of a multitenant building is shown in Figure 10-20. The

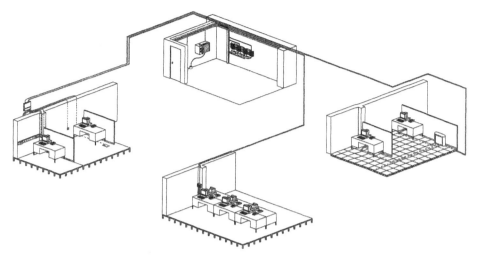

Figure 10-20 Zone distribution cabling. (*Source:* Courtesy of TYCO Electronics Corp.)

plan includes the local telecommunication closet and examples of consolidation, multiuser outlets, under-floor connection, and Pypline® cabling for a modular furniture connection.

10.10 SMALL BUSINESS SCENARIO

Small office-type businesses seldom have the resources to install an expensive LAN or to have an in-house telecommunication support staff. The solution is a system that

- Is cost effective
- Is easy to install
- Requires little maintenance
- Has internal diagnostics for quick problem analysis
- Has connectability to the Internet for individuals
- Has an advanced design that can be updated with future technology
- Can be expanded to a larger network

One solution is that shown for a small law firm in Figure 10-21. The architecture may be twisted pair, coax, fiber, or a combination of these. The wiring media will depend on availability and budget.

A small business may have several sites that need to be integrated into a LAN/WAN for

- Reliable communication between offices and personnel
- Online credit card service
- Connection to the Internet

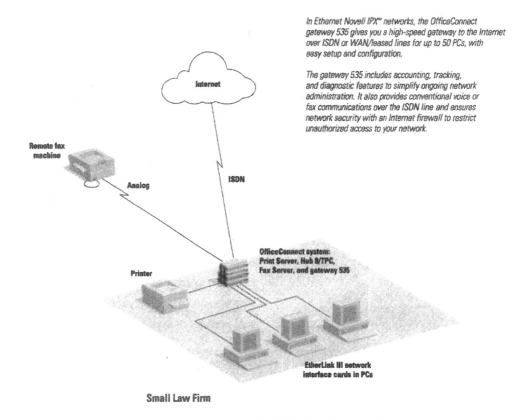

In Ethernet Novell IPX™ networks, the OfficeConnect gateway 535 gives you a high-speed gateway to the Internet over ISDN or WAN/leased lines for up to 50 PCs, with easy setup and configuration.

The gateway 535 includes accounting, tracking, and diagnostic features to simplify ongoing network administration. It also provides conventional voice or fax communications over the ISDN line and ensures network security with an Internet firewall to restrict unauthorized access to your network.

PCs connect to the Hub 8/TPC, which is clipped in a single stack with the OfficeConnect Print and Fax Servers and the gateway 535. These units give all the PCs freedom to use the office printers and desktop fax service.

Figure 10-21   Small office LAN. (*Source:* Courtesy of 3Com Corp.)

The architecture must be cost effective, reliable, easy to maintain, and flexible to accommodate future expansion as the business grows. The business shown in Figure 10-22 is comprised of the main office, the business office, a retail outlet, home-based offices, and the Internet.

The wiring media and connections for a small business are more complex than those for a small office and probably involve a history of wiring and equipment addition as the business grows. Expansion or major changes to a network require a telecommunication professional to implement a plan and follow through on the installation.

The larger the organization, the greater the need for planning to improve or update an existing network or to implement a completely new system. It usually takes more planning to modify an older, possibly outdated wiring system to new standards than to begin from scratch. However, only in newly constructed facilities will the telecommunication professional have this luxury. A large corporation will probably require media comprised of twisted pair, coax, fiber,

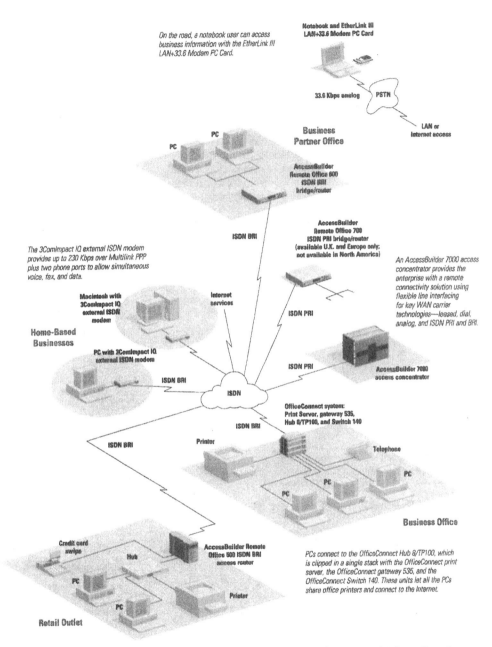

Figure 10-22 Connection for a small retail business. (*Source:* Courtesy of 3Com Corp.)

and wireless. The wireless may consist of microwave or infrared technology. Figure 10-23 depicts a network for such a corporation. The LAN/WAN connects the corporation headquarters, a remote branch office, and an engineering/design facility, with access to the Internet.

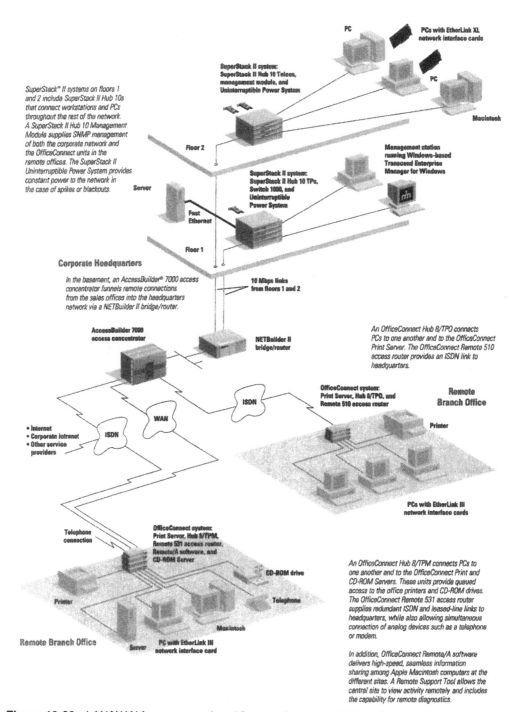

Figure 10-23 LAN/WAN for a corporation. (*Source:* Courtesy of 3Com Corp.)

10.11 SUMMARY

Cabling systems are becoming more complicated and more important as the telecommunication needs of individuals and corporations become greater. Increased transmission speed increases the need for seamless media conduit.

QUESTIONS

1. Make an architectural cabling diagram for the LAN in Figure 10-19. Select the wiring media and list the materials necessary to implement the system.
2. Make a simplified cabling diagram of the LAN in Figure 10-20 showing the cabling media necessary for implementation of media types.
3. Indicate the type of cabling for each connection in the LAN/WAN depicted in Figure 10-23.

CHAPTER 11

Testing and Troubleshooting

11.1 INTRODUCTION

Testing, troubleshooting, and maintaining a communication system is an involved procedure that encompasses both hardware and software specialists. Some of the testing and troubleshooting can be accomplished by the use of a diagnostic software program.

These functions of testing, troubleshooting, and maintaining fall in the lower level of system protocol. For example, cable testers address the wiring and connections in the first four layers of the Open Systems Interconnection (OSI) model developed by the International Standards Organization (ISO).

Layer 1—Physical. The physical means of sending data over the lines. Functional control of data circuits, electrical and mechanical.

Layer 2—Data link. Procedures and protocols for operating the communication lines and detecting and correcting message errors.

Layer 3—Network. Addresses routing between networks and determines how information is conveyed between computers.

Layer 4—Transport. Delivery of information between networks, including flow control and error recovery.

Layer 5—Sessions. Coordination of information to ensure that all information is received.

Layer 6—Presentation. Provides common ground for transferring data between systems by providing transparent communications between systems with different data formats.

Layer 7—Applications. Provides functional applications services such as file access, virtual terminals, remote file access, and so on.

LAN/WAN testers concentrate on Layers 1 through 4, providing comprehensive statistics of error of data frames and packets, lists of top users, and maps of network users by address. Most protocol analyzers can simulate traffic and examine communications on the network at any level of the ISO mode. PC-based network testing units can provide additional information, such as database or report generation, and simultaneously deal with two different protocols. There are network traffic simulators that utilize simulation software that closely simulates real-time applications and ascertains the effectiveness of proposed applications on an old network.

Networks demand a lot of attention. Often network failure can be predicted and corrected before a crash.

11.2 OBJECTIVES OF TESTING AND TROUBLESHOOTING

Testing procedures and methods have three objectives:

1. Train personnel to ensure that all specialists, engineers, and technicians who are required to perform test sequences are qualified on all test equipment. As part of this requirement, each operator should take a "hands-on" test periodically to ensure that correct technical analysis is derived from test equipment data.
2. Train all personnel on new test procedures and test equipment as more efficient and effective equipment becomes available.
3. Locate and identify defective components within the system. This should be accomplished by systematically testing by functional groups. This procedure, sometimes called "bracket-troubleshooting," is helpful in performing corrective maintenance.

The issue of functionality and testability must be addressed for any equipment or cabling system. However, it is more difficult to incorporate the second item of testability into each circuit or system. Most designers now plan in test function devices as part of the system. Other designers provide breakout test points so that test equipment can be "hung on" or switched into the circuit to test it. It is poor design not to address the problem at the offset.

There are three testing access needs to be addressed in system testing: manual access, switched access, and in-line access. In-line access may utilize both manual and automatic testing and analysis methods. The *manual* method usually employs the use of test jacks, test points, and/ or breakout panels of individual components sometimes through test patch panels and terminal strips. The technician will usually need to bring the test equipment to these locations to make the test and determine the source of the problem. Usually such equipment must be small and portable since the test points may be in different locations throughout the system. These access points allow test equipment to be hung on to the circuit at various points to determine functionality. Test points should be placed by the cabling system designer in locations that "bracket" or break the system into logical components. However, often when troubleshooting a problem on a system of prescribed test points, it is necessary to test a subsystem or circuit for which no test points are provided. This method is usually the most labor intensive and least equipment intensive since the same equipment can be used at different locations.

The *switch-standby* troubleshooting method requires that test equipment be positioned near or at the circuit test points. Equipment can be "plugged or switched" into the line or prescribed test points. This method is more equipment intensive than the manual method access but is more efficient because there is less equipment setup and installation time. Test equipment stations are installed at critical points throughout the system. Each station will have the necessary devices to perform various types of analysis based on a predetermined number of test procedures. There will be some equipment redundancy with this method.

The *in-line access (manual analysis)* method uses passive devices that are installed in the circuit. Overall system efficiency is usually slightly degraded by the installation of passive test equipment. However, there are many advantages as the equipment can serve as a "performance monitor" of system and circuit activity. This type of monitoring is especially helpful when introducing new devices into the network. Introduction of new devices can cause a change of transmission characteristics that may degrade the transmitted signals. This method is the most costly and generally employs the most sophisticated equipment.

In-line access (automatic testing and analysis) requires that some in-line devices be used to perform *automatic line testing*. Such devices are computer based and can be programmed to run a variety of tests and functions on an ongoing basis without interfering with network traffic. Once the testing phase has been finished the device will generally generate a test data file. This file can be routed to specialists for review and, if necessary, development of a locate and repair plan.

The data record can also be sent to an "analysis program" for review. Such programs take the raw data (usually numeric), perform calculations with the inputs, and compare the data against predetermined limits. The results of this analysis are then sent on for automatic distribution via a routing list. When the resulting values exceed predetermined limits, a warning or alarm can be given to notify the technical support center of the problem. The following might appear on the monitor screen.

```
          !!! RED FLAG ALERT !!!
1. DOCUMENT AND CORRECT ALL INCORRECT LABELS.
2. DOCUMENT AND CORRECT ALL IMPROPER CABLE RUNS, CONNECTOR-TYPES,
   AND/OR INSTALLATIONS.
3. DOCUMENT AND CORRECT PATCH PANEL, BREAKOUT BOX, AND TERMINAL
   LABELING.
4. UPDATE DATABASE EVERY TIME A CORRECTION WAS INITIATED OR
   PERFORMED.
```

Be sure to look for any "RED FLAG ALERTS" during your work activities. If any are observed, correct as necessary or initiate appropriate paperwork to do so.

Testing of each type of wiring medium described in Chapters 2 through 4 may be somewhat different, but the purpose of testing is the same in each case: to provide a medium over which to transport voice or data information with a minimum amount of loss, distortion, noise pick-up, and cross talk. We will examine methods used to test each type of wiring medium: twisted-pair cable, coaxial cable, and fiber-optic cable.

11.3 TESTING TWISTED-PAIR CABLES

Any time cables are installed or there is a cabling change for installation of new equipment, the installation and testing of twisted-pair cables should include the following:

1. The correct labeling of all wires and all termination points. Wire labeling may, in small installations, be by coloring of the wiring insulation and writing the code on a wiring list. However, each termination point must be labeled to identify it on a wiring list. Labeling at termination points or terminal connections should identify the difference between telephone and data lines.

2. The testing of each wire within an installation for continuity, opens, and shorts between wires and shorts to ground.

3. Polarity testing to assure that the signals from a source will arrive at the destination with the proper polarity.

4. Labeling all lines that are operating in the DSL mode so that they may be tested for audio, data, and video.

Any time there is a cabling change entailing addition of cables, upgrading, equipment replacement, or relocation of equipment, every cable must be tested for continuity or a short circuit and every connector must be checked to ensure that it is wired correctly. This may seem like an unnecessary procedure and a complete waste of time, and 99% of the time it is. However, that one percentage could cause days of troubleshooting of the system to find a short or open circuit after the system is complete. It is also important that any cable changes be entered in the documentation table.

11.4 CONTINUITY TEST OF A CABLE

A continuity test assures that a cable is complete from end to end. This test is accomplished with an ohmmeter in volt-ohms-millimeter (VOM) or in a digital voltmeter (DVM). A DVM used on the diode scale will sound a tone when a short circuit or continuity is found. A device called a *continuity tester* can also be used for the tests of twisted-pair cables. Continuity testers are simple ohmmeters that indicate a short or continuity by a tone or a light.

The VOM shown in Figure 11-1a has an analog scale. When the test leads are shorted together, the meter pointer will read zero on the ohm scale, as shown in Figure 11-2a. When the ohm function is selected on the DVM the readout will indicate zero ohm (Figure 11-2b).

When the test leads are shorted together on the continuity checker, the light will be illuminated. More expensive continuity testers will sound an audible alarm when continuity is complete. When the VOM is used for a continuity or short test, the test leads should be shorted and

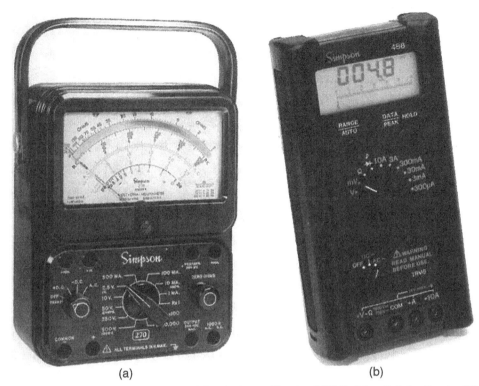

Figure 11-1 Basic test instruments: (a) volt-ohm-millimeter (VOM), (b) a digital voltmeter (DVM).

the pointer adjusted to the zero position with the ohm adjustment. This adjustment compensates for aging of the internal battery in the VOM.

To determine the continuity of a short cable, the VOM or DVM can be connected to a wire on one end of the cable and the other lead of the instrument connected to each wire at the other end of the cable until a zero response is indicated (Figure 11-3). When the wire is found, the pointer on the VOM will indicate zero or the readout on the DVM will indicate zero or near zero ohm, depending on the resistance of the wire. When an indicator is used, the light will indicate continuity. An open wire would result in no indication on either of the instruments. A short to ground (grounded wire) could be found by connecting one lead of either instrument to ground and then connecting the other lead to one wire at a time (Figure 11-4).

11.5 A SHORT TO GROUND TEST

A shorted wire to ground would give the same indication on the instruments as wire continuity. A partial short will indicate an undetermined resistance reading.

Each wire should be tested against every other wire to ensure that no two wires are shorted to each other. This is accomplished by connecting one lead of the continuity/short testing instru-

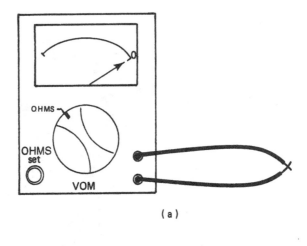

(a)

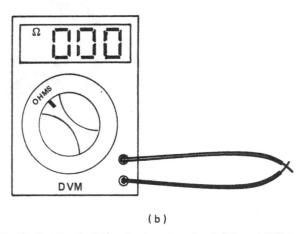

(b)

Figure 11-2   Setting the OHMS adjustment on the VOM and DVM.

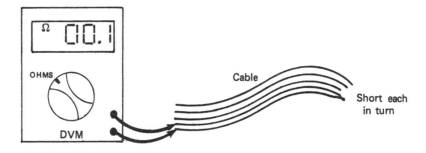

Figure 11-3   Use of the ohmmeter to test for continuity on a short cable.

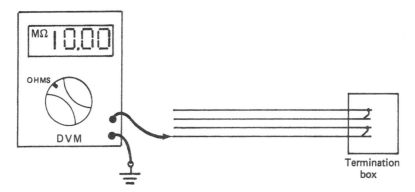

Figure 11-4 Testing of a long cable with a VOM, DVM, or continuity tester.

ment to one wire and then connecting the other lead to each wire in turn. The process is repeated until all wires are tested to every other wire. A record of each wire test should be recorded to simplify the test. For example, in an 8-wire cable, wire 1 would require 7 test, wire 2 would require 6 test, wire 3 would require 5 test, and so on.

The testing of a long cable is somewhat more difficult in that it is impossible to connect the testing instrument lead to both ends of the cable. However, any two wires can be clipped together at the end away from the test instrument and the resistance of both wires read at the other end (Figure 11-5a). The resistance of the wires should be very low (see Table 11-1) and can be estimated (if the length and wire size are known). The wires must be clipped to the first wire one at a time and tested. The resistance of any pair of wires should be low and be almost the same value. If the cable is very long, two people can perform the test, communicating with a two-way radio or by telephone.

A ground wire test on a long cable is performed in much the same way, with one lead of the test instrument connected to a ground that is common to both ends of the cable (Figure 11-5b). The wires are then connected to the continuity instrument one at a time. The other end of the wires must be open. This test can be performed by one person. Only those wires that are purposely connected to ground should show a response. A short test between wires would be performed by one person in the same manner as was described for a short cable.

11.6 TONE TEST GENERATORS

Many voice and data network problems are cable related. Troubles can often be traced to miswired connections, loose connections, intermittent shorts or opens, and open or shorted pairs. These problems can be located with the use of a tone generator and pick-up amplifier. A tone test generator can be used to apply a tone to a single conductor, to a conductor pair, to a coaxial cable, or even to a deenergized AC electrical line. An inductive amplifier probe can then be used to locate that conductor within a bundle at a cross-over point or at the remote end by pressing the probe against the injected wire. An example of a tone generator and pick-up probe is shown in Figure 11-6.

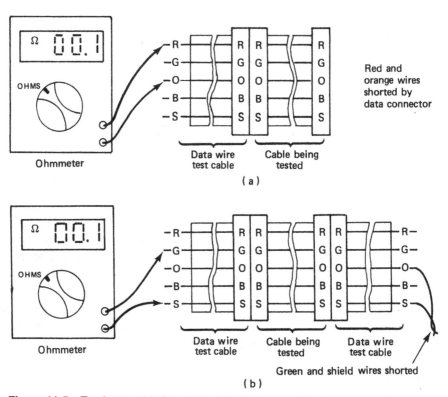

Red and
orange wires
shorted by
data connector

Ohmmeter

Data wire
test cable

Cable being
tested

(a)

Ohmmeter

Data wire
test cable

Cable being
tested

Data wire
test cable

Green and shield wires shorted

(b)

Figure 11-5 Testing a cable for shorts: (a) wire to wire, (b) wire to ground.

Table 11-1 Wire resistance

Gage AWG or B&S	Diameter		Circular	Resistance	Weight
	(in)	(mm)	mil area	@ 68° F Ω/1000 ft	1000 ft
20	.0320	.813	1020.0	10.15	3.092
21	.0285	.724	724.0	12.80	2.452
22	.0253	.643	640.0	16.14	1.945
23	.0226	.574	511.5	20.36	1.542
24	.0201	.511	404.0	25.67	1.223

AWG = American wire gauge; B&S = Brown and Sharpe

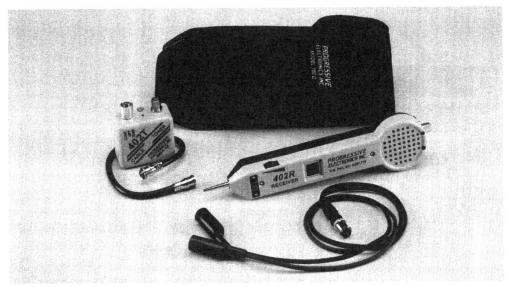

Figure 11-6 Tone generator and inductive amplifier. (*Source:* Courtesy of Progressive Electronics)

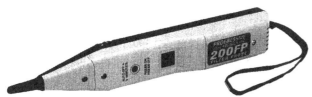

Figure 11-7 Filter pick-up probe. (*Source:* Courtesy of Progressive Electronics)

In some cases the tone signal may be lost in background noise of 60 Hz through electromagnetic pick-up from power lines or electronic devices. The pick-up probe in Figure 11-7 has an internal filter to remove such interference. It is sometimes more convenient to test wiring through a modular plug or jack. The tester shown in Figure 11-8 can be used to find opens, shorts, reversals, and transposed pairs, and to identify an active circuit. More expensive tone test instruments, with sufficient output power, can be used to locate wires in walls or underground. Figure 11-9 illustrates the use of such an instrument to locate underground cables. The device can locate cable up to 3 ft underground and be used to determine the approximate depth of cable before digging.

11.7 TESTING A COAXIAL CABLE

A continuity test of a short coaxial cable can be performed in the same manner as for a short twisted-wire cable. The ground should be checked from end to end for continuity. A signal lead to ground test should also be performed on every coaxial cable.

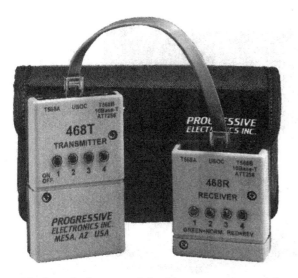

Figure 11-8 Modular cable tester. (*Source:* Courtesy of Progressive Electronics)

Testing of a long coaxial cable presents a special problem in that there is one signal wire and one ground wire. If we short the signal wire to the only other lead (the ground), a continuity test will indicate the same as a shorted cable. For this reason a ground test should be performed first to be sure that there is an open between the signal lead and the ground lead.

When the technician is assured that the cable has no internal short between the signal lead and ground, a continuity test can be performed. This is accomplished by shorting the signal to the ground leads (Figure 11-10) and measuring between the conductors at the other end of the cable. Continuity is indicated for both the ground shield and the center conductor. These tests are very important for both preassembled cables and those custom made by the technician. A manufactured cable may look perfect, but a factory assembly can be faulty.

A tone generator is usually equipped with BNC, TNC, or twisted-pair input connector for convenience. Some versions have a mounted light to indicate a visual check of continuity or short circuit as well as a speaker for an audio tone indication. An additional device that can greatly assist the wiring specialist is a "shorting cap" (Figure 11-11a), a male or female cable connector that has the center conductor of the coax shorted to the shield or ground. In this manner the device serves to short one end of the cable. This small device can be connected into the distant end of a cable and the technician can go to the near end and proceed to check for cable identity. With proper equipment the cable can be tested for length, impedance, open/short condition, and so on.

A shorting cap can be modified with the addition of a light emitting diode (LED) across the center conductor to ground (Figure 11-11b). This device can be used to indicate the presence of a low voltage on the line. This device is also useful to determine "hot" controller ports or PC/workstation connections. It can also be used to check line continuity in conjunction with a tone generator. This is accomplished by placing the tone generator on one end of the line and the modified cap on the other end. This procedure performs a simple voltage line test. If the LED

Figure 11-9 Locating an underground cable. (*Source:* Courtesy of Progressive Electronics)

glows brightly, the line resistance is low and the battery-powered tone generator is operating effectively. However, if the LED glows dimly, the battery of the generator is weak or the line has a high resistance. If the LED shows no light, the generator is faulty or the line is open.

The tone generator and inductive amplifier discussed previously are useful in testing coaxial cable. Figure 11-12 shows a Thinnet cable tester with BNC connectors. The unit can be used to identify shorts, opens, or terminated cables.

Determining that copper wire is open or shorted and repairing the fault is usually the simplest part of the repair job. Finding the location of the trouble can be very difficult. A long cable may run underground, in walls, or between buildings. Cabling blueprints should show the rout-

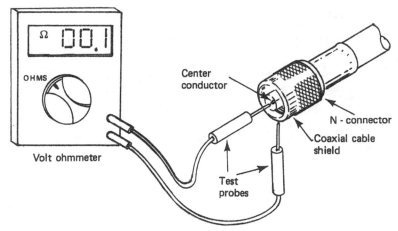

Figure 11-10   Testing for internal shorts with an ohmmeter.

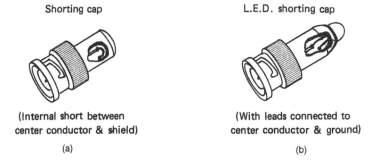

(Internal short between   (With leads connected to
center conductor & shield)   center conductor & ground)

(a)   (b)

Figure 11-11   A shorting device for testing coaxial cable: (a) shorting cap,
(b) LED shorting cap.

ing of the cable. The distance of the break or short can be determined by the use of a resistance
fault meter such as shown in Figure 11-13. Figure 11-13b depicts the measurement of copper
wire through a registered jack (RJ) telephone jack.

11.8 STANDARDIZED TESTING PROCEDURES AND ONLINE DATABASE

To provide uniform test and timely repair operations, a series of standardized testing procedures
(STPs) must be agreed on and developed. These procedures need to be established in two different forms. First, an *online procedure database* should be established and maintained. Second, a
series of *standard procedures manuals* need to be installed and maintained at each work area
and/or technical workstation (as appropriate).

The information to create this document, including procedures and techniques, is obtained
from the device-use and device-training manuals, product classes, seminars given on problem

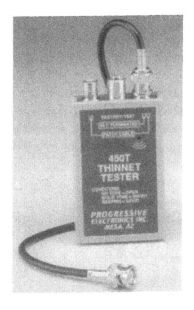

Figure 11-12 Thinnet cable tester. (*Source:* Courtesy of Progressive Electronics)

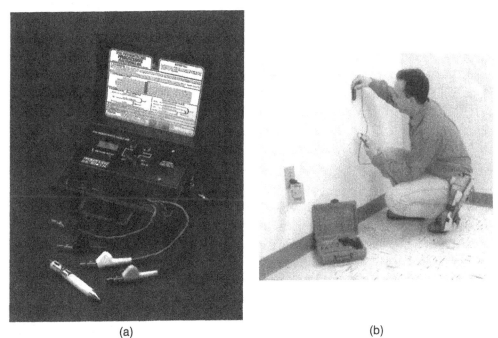

(a)   (b)

Figure 11-13 Resistance fault meter: (a) instrument, (b) application. (*Source:* Courtesy of Progressive Electronics)

determination, and personnel of each department. Engineers and specialists with hands-on and/ or academic experience should be asked to input their knowledge to the manual.

The contents of these manuals may include installation and testing procedures, trouble- shooting guidelines, instructions for installing microcode on a controller and downloading a pro- gram from the host to perform a special test on a workstation, and so on. Once all procedure testing outlines are placed in the reference manual and a complete set is agreed on, an accep- tance test should be run. To run an acceptance test, a technician is placed into a testing bay with simulated (or real) symptoms and problems. He or she then uses the procedures step by step to isolate the problem and determine the cause or the next course of action.

There should be a three-level evaluation of the test method and applications:

1. Ability to read and comprehend the text
2. Ability to perform the testing task indicated
3. Ability to evaluate the test results to determine accurately the problem cause

These procedures and guidelines should be documented in two ways:

1. A series of departmental reference books should cover the major technical activities that the department is charged with. These manuals should outline what needs to be done at each level and station within the department. The organization should maintain a manual at each work area.
2. A test and troubleshooting database must be built and maintained. This database should outline each work activity and test procedure, including installation, corporate standard- ization layouts, troubleshooting procedures, escalation guidelines, and carry-over and sup- port contacts for additional assistance. This online database should be available on a 24- hour basis to all departmental employees.

The advantage of this system is that for the specialist in the field, an online system can be a valuable resource to look up test or troubleshooting procedures and to inquire about the next level of management contact escalation. Whoever is charged with maintaining the database must also ensure that a set of reference volumes exists within the organization. This individual should be responsible for maintaining and updating these manuals.

The final test and repair manual should be placed in all work areas for wiring media tech- nicians' use. If technicians are unsure of how to proceed or unclear about the next step in trou- bleshooting a problem, the procedures manual should provide direction and guidance.

Finally, all system and wiring problems should be logged in a maintenance log with the date of occurrence, the symptoms observed, and the corrective procedures. This log will often give a quick answer to network problems, as many problems tend to be repetitive.

11.9 TROUBLESHOOTING CABLE WITH A TIME DOMAIN REFLECTROMETER

Time domain reflectrometers (TDRs) can be used to locate cable problems in any metallic two-conductor cable, including the following:

- Map and document cable systems
- Detect thief of service
- Locate shorts, opens, and partial opens
- Capture intermittent faults
- Find splices and resplices
- Locate bridged taps
- Locate ground faults
- Locate water damage
- Locate crimps
- Locate cuts in the cable
- Locate smashed, pinched, or kinked cable
- Locate shorted conductors
- Locate and identify system components
- Determine cable length
- Clear lines for ADSL, ISDN, and/or HDSL
- Find splits and resplits
- Locate problems caused by construction
- Verify cable installation before acceptance
- Measure dBRL (reflection level) of the fault
- Measure length of new or old cable reels for breaks or splicing

The TDR transmits a signal pulse down a wire at a certain speed, and at the point of a problem some of the signal is reflected back to the unit. If the speed of the signal is known, and the time between transmission of the reflected signal is divided by two, the distance to the problem can be determined. In this regard, the TDR operates as radar where a high-frequency pulse is transmitted into the air and a small amount of the signal is reflected back to the receiver from a target. In free air the signal travels at approximately the speed of light—186,000 miles or 300×10^6 m per second. For example, if an airport tower radar transmitted a signal over the ocean and the signal returned in 1080 µs, the distance would be

$$Distance = \frac{186000 \text{ mps}}{2} \text{time delay} = 93,000 \times 1080 \times 10^{-6} = 100 \text{ miles}$$

The speed of the signal down a wire depends on the type of wire and is never at the speed of light. The ratio of the speed that a signal travels down a cable to the speed of light in a vacuum is called the cable's *velocity of propagation* (VOP). The speed of light in a vacuum is 186,000 miles per second or 300,000 km per second. This speed represents the number of 100%. A coaxial cable with a VOP of 0.85 would transmit a signal at 85% the speed of light. Each type of wire has a VOP determined by the wire type, insulation, and other physical characteristics. The VOP of a wire must be considered when testing with a TDR. Table 11-2 shows the VOPs of different types of cable. The preceding example of the radar would have to be modified by a factor of 0.67 if the signal were transmitted down AWG 24 Teflon-coated telephone wire. The VOP is the percentage of the speed of light that a signal travels down a cable. For example, the aforementioned

Table 11-2   Cable types and their VOPs

Telephone	Pic		19 Gauge	.912 MM	.72	Polyethylene	.66	
			22 Gauge	.643 MM	.67	Polypropylene	.66	
			24 Gauge	.511 MM	.66	Teflon	.67	
			26 Gauge	.404 MM	.64			
	Jelly Filled		19 Gauge	.912 MM	.68			
			22 Gauge	.643 MM	.62			
			24 Gauge	.511 MM	.60			
			26 Gauge	.404 MM	.58			
	Pulp		22 Gauge	.643 MM	.67			
			24 Gauge	.511 MM	.68			
			26 Gauge	.404 MM	.66			
CATV	Belden (foam)			.78 - .82		Capscan (foam)	.82	
	(solid)			.66		CC	.88	
	Comm/Scope (F)			.82		CZ Labs (foam)	.82	
	Para I			.82		General Cable		
	Para III			.87		RG-59	.82	
	QR			.88		MC²	.93	
	Times Fiber RG-59			.93		Scientific Atlanta		
	T4, 6, 10, TR+			.87		RG-59	.81	
	TX, TX10			.89		Trunk	.87	
	Dynafoam			.90				
	Trilogy (F)			.83				
	7 Series			.88				

Power	Impregnated paper		150–171 (.50 – .57)	PVC	152–175 (.51 – .58)
(European)	Dry paper		216–264 (.72 – .88)	PTFE	approx. 213 (.71)
	PE		approx. 200 (.66)	Air	approx. 282 (.94)
	XLPE		156–174 (.52 – .58)		

Power	XLPE	345	35	1/0	.57	XLPE		15	#4 CU	.52
(U.S.)	XLPE		35	750 MCM	.51	XLPE		15	500 MCM	.53
	PILC		35	750 MCM	.52	XLPE		15	750 MCM	.56
	XLPE		25	1/0	.56	XLPE	260	15	750 MCM & AL	.53
	XLPE	260	25	1/0	.51	EPR	220	15	1/0	.52
	XLPE		25	#1CU	.49	EPR	220	15	4/0	.58
	PILC		25	4/0	.54	EPR		15	#2 AL	.55
	XLPE	175	15	1/0 AL	.55	PILC		15	4/0	.49
	XLPE	175	15	1/0	.51	EPR		5	#2	.45
	XLPE		15	2/0	.49	EPR		5	#6	.57
	XLPE		15	4/0	.49	XLPE		.6	1/0	.62
	XLPE		15	#1 CU	.56	XLPE		.6	4/0	.62
	XLPE		15	#2 CU & AL	.52	XLPE		.6	#2	.61
	XLPE		15	#2 AL	.53	XLPE		.6	#8	.61
	XLPE		15	#2 AL	.48	XLPE		.6	#12-6PR	.62

Table 11-2   Cable types and their VOPs (Continued)

LAN	UTP 26		.64	IBM	1	.64		
	Thinnet		.66 – .70		2	.66		
	Ethernet		.77		3	.70		
	Token Ring		.78		4	.72		
	Arcnet		.84		5	.76		
	Twinaxial Air		.80		6	.78		
	Twinaxial		.71		7	.82		
	Appletalk		.68		8	.84		
					9	.82		
Land/	Andrew			Cablewave				
Mobile	Radiax	All	.79	FLC 12-50J			$1/_2$"	.88
				FLC 78-50J			$7/_8$"	.88
	Heliax							
	FHJ 1-50	$1/_4$"	.79	Cellflex FoamFCC + FLC				
	FSJ 1-50	$1/_4$"	.78	FCC 38-50J			$3/_8$"	.81
	FSJ 4-50B	$1/_2$"	.81	FLC 12-50J			$1/_2$"	.88
	LDF 2-50	$3/_8$"	.88	FLC 78-50J			$7/_8$"	.88
	LDF 4-50A	$1/_2$"	.80	FLC 158-50J			$1 \, 5/_8$"	.88
	LDF 4-75	$1/_2$"	.88					
	LDF 5-50A	$7/_8$"	.89	Celwave		All		.88
	LDF 7-50	$1 \, 5/_8$"	.88					
	FT 4-50	$1/_2$"	.85	Coax Transmission Line				
	FT 5-50	$7/_8$"	.89	920213			$7/_8$"	.99
	HJ 4-50	$1/_2$"	.91	920214			$1 \, 5/_8$"	.99
	HJ 5-50	$7/_8$"	.92					
	HJ 5-75	$7/_8$"	.90	Flexwell HCC				
	HJ 7-50A	$1 \, 5/_8$"	.92	HCC 12-50J			$1/_2$"	.91
	HJ 8-50B	3"	.93	HCC 78-50J			$7/_8$"	.91
	HJ 11-50	4"	.92	HCC 58-50J			$1 \, 5/_8$"	.95
	HJ 9-50	5"	.93	HCC 300-50J			3"	.96
				HCC 312-50J			$3 \, 1/_2$"	.96
				HF 4 $1/_2$ CU24 4			$1/_2$"	.97

Source: Courtesy of Riser Bond Instruments

0.67 factor for twisted pair means that the signal travels down the wire at a rate of 67% the speed of light.

Figure 11-14 depicts three typical TDRs. The unit in Figure 11-14a has an analog readout, while the unit in Figure 11-14b has a digital readout, and the unit in Figure 11-14c is hand-held. Figure 11-15 depicts a TDR in use at a twisted-pair junction point.

The width of the signal pulse limits the minimum distance to which a fault can be measured. During pulse time the receiver is "blanked out," creating a blind spot. For this reason any test should begin using the shortest pulse width. If the fault is not located, the pulse width should be increased progressively; if the fault is still not located it is probably in the blind spot. In this case the fault can be located by adding a known length of cable and retesting. The added cable length must be subtracted from the measured distance to the fault. The added cable, if possible, should be the same type as the cable being tested so that the propagation velocity is the same. Table 11-3 gives the blind spot distances for various pulse widths.

The wider the pulse width, the more energy that is transmitted down the cable and the easier it is to see the fault. The attenuation of the signal by the cable and a small fault may require the wide pulse to reflect enough energy to be sensed by the TDR.

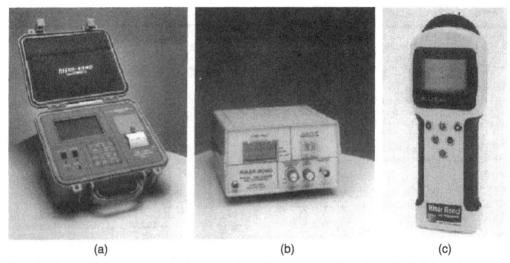

(a)   (b)   (c)

Figure 11-14 Time domain reflectrometer: (a) with analog readout, (b) with digital readout, (c) with hand-held instrument. (*Source:* Courtesy of Riser Bond Instruments)

Figure 11-15 A TDR being used to test cables at a twisted-pair junction point.

As we noted earlier, the VOP of a cable must be known to make an accurate measurement with a TDR. The VOP may be found in the manufacturer's information on the packaging, through installation data in the cable log, by contacting the manufacturer, or by measuring the VOP on the same type of cable of known length.

Table 11-3 Blind spot distance

Pulse Width	Blind Spot			
	Twisted Pair		Coaxial	
sub-ns	1	0.1 m	1 ft	0.1 m
2 ns	6 ft	2 m	6 ft	2 m
10 ns	12 ft	4 m	14 ft	4 m
100 ns	50 ft	16 m	55 ft	17 m
1 μs	400 ft	120 m	430 ft	133 m
2 μs	630 ft	192 m	850 ft	360 m
4 μs	1300 ft	390 m	1600 ft	487 m
6 μs	1970 ft	600 m	—	—

A sample test cable should be as long as possible, at least 33 m or longer. When measuring the VOP of a cable, keep in mind that the value can change with temperature, humidity, and age. The VOP of new cable will vary slightly from the manufacturer's data.

Cable loss/return loss is the ratio in decibels of the signal transmitted down a cable to the signal returned. When a signal is transmitted along a cable, some of the energy is lost due to resistance and impedance of the cable. This cable loss is measured in decibels. When the transmitted signal energy reaches a fault, some of the signal is returned. The greater the fault, the larger the returned signal. The ratio of the signal transmitted to that returned is called the return loss and is measured in *decibel return loss* (dBRL). Return loss is a way of measuring the impedance of a cable. A large return loss means that most of the signal was not reflected but continued down the cable. A small return loss means that the energy was dissipated in a fault and not returned. A completely open or short would return no energy and therefore result in 0 dBRL. Figure 11-16 depicts the results of the two foregoing examples.

Sometimes a cable can contain multiple faults that can be caused by a number of factors. If a fault is a total short, the TDR will read only that fault. However, if the fault is a partial short, the TDR will read other faults. Figure 11-17 depicts a test that indicates two faults in a cable. The first fault produces a large energy reflection, while the second fault produces a much smaller energy reflection. This is because of cable attenuation and signal reduction by the first fault. The four pulses on the returned wave are the transmitted pulse (blind spot), the first major fault, the second small fault, and the reflected signal from the open end of the cable.

It is desirable to open the far end of a cable that is being tested as the cable termination will absorb much of the signal energy. Sometimes this is not feasible and a fault can still be found if the cable is damaged, as a signal will be reflected prior to absorption by the termination. If the termination is connected and a signal is returned from that point, the termination may be faulty.

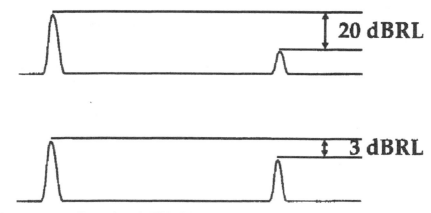

Figure 11-16 Examples of dBRL: (a) a high impedance resulting in a low dBRL, (b) low cable losses resulting in a high dBRL.

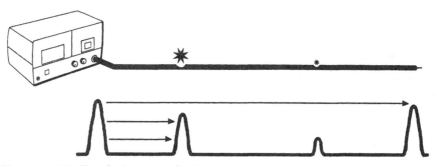

Figure 11-17 Two fault signal reflections: (a) instrumentation connection, (b) transmitted and reflected signals. (*Source:* Courtesy of Riser Bond Instruments)

For accuracy in locating a fault, the cable should be tested from both ends. This eliminates error in the case of multiple faults. It also gives greater exactitude in locating a fault error that might be caused by an incorrect VOP or factors that change the VOP of a cable.

The connection between a cable being tested and the TDR is very important to minimize test signal loss and ensure accuracy of measurements. The connections should connect directly to the input jacks of the TDR and be the same type as the cable. For example, a 75-Ω coaxial cable should be connected with a 75-Ω connector. Figure 11-18 illustrates examples of correct and incorrect connection methods.

The TDR is a useful and accurate instrument. Most errors are "cockpit errors" caused by the user. It is very important that the wiring technician be completely familiar with the instrument before performing a test. Some general rules are as follows:

• Be aware of the operating theory of the TDR. Test the instrument's operation on a known cable, learn the function of each key, and identify waveform signatures.

Correct Connection Incorrect Connection

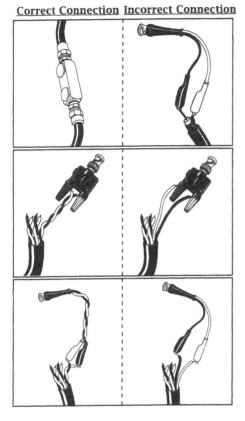

Figure 11-18 Good and bad cable connections. (*Source:* Courtesy of Riser Bond Instruments)

- Read and follow the manual's operating instructions. Read the operator's manual and review any instructional material before a failure happens.
- Make a good connection between the TDR and the cable under test. Whenever possible, connect directly to the instrument's front panel.
- Get as close to the fault as possible. Isolate the trouble to the smallest length of cable possible.
- Enter the correct VOP for the cable under test. By entering the correct VOP you are matching the test instrument to the cable.
- Start with the shortest pulse and work up. The pulse blanks the instrument for a given time. A fault that is within the blanking could be masked from view. Increase the pulse width in increments. If a fault does not appear, add a known length of cable between the instrument and the cable under test.
- Test from both known lengths of cable between the instrument and the cable under test. This will assure the correct location of the fault.
- Determine the cable's path and/or depth. The distance reading will be for the cable and will not take into consideration curves, loops, and drops.

Figure 11-19 Hand-held time domain reflectometer.

- Always retest a repaired cable to ensure that the repair is effective and that there is no other fault.
- Keep notes and make diagrams as the cable is being tested. You may be repeating this test at a later day.
- Record faults and correction in the cable log. Faults have a way of repeating themselves. Good bookkeeping may save time in the future.
- Use logic. Identify events that happened before the cable crash. For example, did a backhoe dig a trench near a buried cable, was a wall moved in an earthquake, or was a modification made to a wall through which the cable passes? Always start at the simplest and cheapest possible of several solutions. In other words KISS—"keep it simple, stupid."

While optical time domain reflectrometers provide a full range of test and restoration functions for cable systems, they are expensive. The cost usually limits the number that are available for servicing on a network to one or two highly skilled technicians. Figure 11-19 depicts a simple hand-held reflectrometer.

11.10 TESTING A FIBER-OPTIC CABLE

The testing of fiber-optic cable offers a very different challenge to the technician. Continuity, open, ground, and short test with VOM or an electrical signal source cannot be performed because the glass fiber is an insulation and as such will not conduct an electric current. The continuity of the fiber can be tested with a light source. However, the efficiency of fiber must be tested with an *optical time domain reflectrometer* (OTDR), which determines the loss in decibels

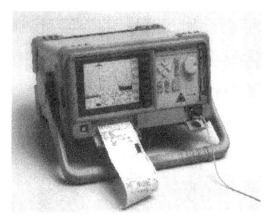

Figure 11-20 Optical time domain reflectrometer. (*Source:* Courtesy of Tektronix Inc.)

of the light signal and the length of the cable by measuring the time that it takes the light to travel end to end. An example of an OTDR tester is shown in Figure 11-20. The OTDR can be used to determine the overall efficiency of the cable system and any fiber-to-fiber connections within the cable.

OTDRs, like TDRs for copper media, can show the presence of splices, connections, and impedance mismatches within the fiber cable. An OTDR determines the loss in decibels of the light signal and the length of the cable by measuring the time that it takes the light to travel end to end. The light will travel through the cable at slightly less than the speed of light (186,000 mi/s or 300,000 km/s). Manufacturer's data or the cable log should be consulted for the light transmission data.

There are three types of tests for a completed fiber-optics loop: continuity testing, link loss testing, and OTDR testing. Link testing and continuity testing are basically the same test and should be completed on all links for certification. Continuity tests mean that the loop is complete and that light passes to the correct termination point. Link loss testing measures the attenuation or loss of light through the cable. This is accomplished by passing a calibrated light in at the source and measuring the power level of the light at the other end of the cable with an optical power meter. The OTDR test measures loss of light on the link and also at which components the loss occurs, resulting in a cable signature. Software such as MS Windows OTDR Emulation will print laser-quality output for record certification.

An optical fiber cable is tested for power loss by a procedure specified as EIA standard FOTP-17, which is depicted in Figure 11-21. The test procedure is as follows:

• Attach the launch cable to the optical source.
• Connect power to the launch cable to measure the output of the launch cable.
• Set an adjustable source to 0 dBm or to –30 dBm. If the source is fixed, record the power reading for later use.
• Connect the launch cable to the test cable with a splice bushing.

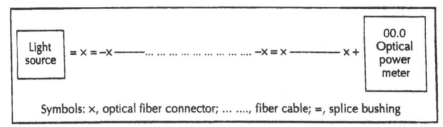

Figure 11-21   Testing an optical cable for power loss.

                    (a)   (b)

Figure 11-22   Instruments used to test fiber-optic cable: (a) light source, (b) power
meter. (*Source:* Courtesy of Corning Cable Systems)

• Connect the power meter to the output of the cable under test and record the power reading.
• Subtract the power meter reading from the recorded launch reading to find the cable power
  loss.
• Compare the cable power loss to allowable loss specifications.

Figure 11-22 depicts examples of a power source and power meter.

11.11 END-TO-END ATTENUATION TEST

End-to-end testing of a fiber cable should be accomplished by using a light source and power
meter in three steps:

Step 1—Connect a short jumper between the power meter and the light source and record
the power meter reading as the reference $Power_{reference}$ in dBm (Figure 11-23a). This is
the output power of the light source into the meter. Never disconnect the jumper from the
light source after this test, as the reference value can change and cause inaccurate final test
results.

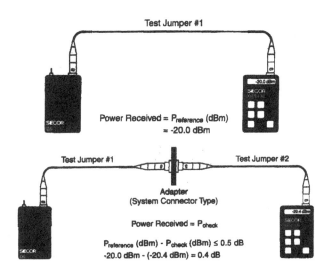

Power Received = P$_{reference}$ (dBm)
= -20.0 dBm

Power Received = P$_{check}$

P$_{reference}$ (dBm) - P$_{check}$ (dBm) ≤ 0.5 dB

-20.0 dBm - (-20.4 dBm) = 0.4 dB

Figure 11-23 Testing the test jumpers. (*Source:* Courtesy of Corning Cable Systems)

Step 2—Disconnect the jumper from the power meter and insert the other jumper that will be used to connect to the cable to be tested, as shown in Figure 11-23b. A typical reading for this Power$_{check}$ test is 0.5 dB of P$_{reference}$. If this criterion is met, proceed to step 3 for the cable test. However, if the reading is greater than 0.5 dB, clean all connections and replace the adapter between the jumpers until the correct result is obtained.

Step 3—Remove the adapter, but leave the two jumpers attached to the power meter and optical source. Attach the detached jumper of the optical source to one end of the cable to be tested and the jumper of the power meter to the other end of the cable to be tested, and turn on the instruments. Record the power level reading in dBm as Power$_{test}$. Repeat step 3 for all fibers to be tested (Figure 11-24a). Some power meters indicate power levels in decibels by storing the reference level and indicating the loss results, as shown in Figure 11-24b.

An optical light source can be used for a continuity test of fiber by applying light to one end of the cable and observing the output at the other end. Figure 11-25 depicts two methods of continuity testing of a coil of fiber cable. Figure 11-25a shows a visible light source connected to one end of a multimode cable with a fiber adapter. The light will be visible at the output end of the reel if the cable is complete. Figure 11-25b shows a laser spliced to one end of a single-mode cable to test for continuity. Again, the light will appear at the output of a complete cable. The laser must be used on the single-mode fiber because the small core requires that a very strong light be used to produce a visible output. Note that the laser can also be used to test multimode cable.

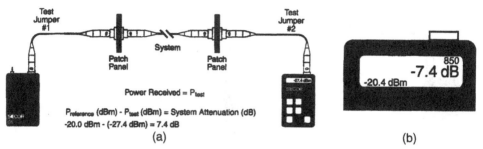

Figure 11-24 Testing a fiber cable: (a) instrument connection, (b) meter reading. (*Source:* Courtesy of Corning Cable Systems)

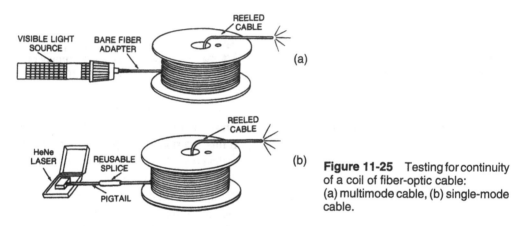

Figure 11-25 Testing for continuity of a coil of fiber-optic cable: (a) multimode cable, (b) single-mode cable.

11.12 CUSTOM-ASSEMBLED CABLES

Communication cable must often be assembled on site by the wiring technician. Long cables may be installed in place to ensure the correct length and then terminated. Great care must be taken in the termination of all three types of communication cable: fiber optic, twisted pair, and coaxial.

Chapters 2 and 3 should be consulted for the assembly of twisted-pair and coaxial terminals. Detailed instructions for terminating fiber-optic cable were given in Chapter 4.

Regardless of the type of cable or the type of connector placed on the cable, a complete continuity and ground test must be performed after assembly, and the connector should be double checked to assure that each wire or fiber is connected to the correct pin of the connector.

11.13 SIGNAL-TO-NOISE RATIO MEASUREMENTS

The signal-to-noise ratio (*S/N*) is a communication measurement that indicates the ability of a circuit or electronic system to distinguish between noise (unwanted signals such as cross talk, electromagnetic pick-up, etc.) and the selected signal. This ratio is measured in decibels (see Chapter 1), and a high decibel reading is desirable since this indicates the receiver's ability to receive a usable signal.

A signal-to-noise ratio of 23 to 26 dB is normal on telephone leased lines. This means that the signal to noise is a power ratio of

$$dB = 10 \log S/N$$

where the S/N is the power of the signal/power of noise. For a low 23-dB ratio the power ratio is

$$23/10 = \log S/N = 199.5$$

and for 26 dB the ratio is

$$26/10 = \log S/N = 398$$

The worst-case signal-to-noise ratio of 23 dB represents a power ratio of signal power to noise power of approximately 200, and the high S/N ratio of 26 dB represents a signal power to noise power of 400. We might note that the 3-dB difference of the S/N ratio represents double the power. An increase of 3 dB at any level represents a doubling of power, while a power reduction of 3 dB at any level represents a 50% power loss.

Table 11-4 presents a sample of decibel loss values versus power loss values. The loss of power in a system is denoted as a minus decibel ($-$dB). We refer to all losses and gains in an electrical circuit as *gains* to simplify the calculation of system gain. For example, suppose that a system had gains of 26 dB and 24 dB and losses of -3 dB and -4 dB. The overall gain of the system would be calculated as follows:

$$\text{System gain} = 24 \text{ dB} + 26 \text{ dB} - 3 \text{ dB} - 4 \text{ dB} = 43 \text{ dB}$$

Table 11-4   A sample of S/N ratios as they relate signal power to noise power

S/N Ratio (dB)	Signal/Noise Power
20	100
21	126
22	158
23	199
24	251
25	316
26	398
27	501
28	631
29	794
30	1000

11.14 REFERENCE POINT FOR POWER LEVEL

Early in the development of the field of communication it was recognized that a *universal reference point* for power level was desirable to which a system power level of gain or loss could be compared. Telephone companies chose to use 1 milliwatt (mW), the approximate output level of conversation from a telephone transmitter, as the reference. This reference level taken across 600 Ω is called a dBm (decibels per 1 milliwatt) and is formulated as follows:

$$dBm = 10 \log \text{signal power in milliwatts/1 milliwatt}$$

The establishment of the dBm, a + or − dB reading on a voltmeter scale, with a value of 1 mW is inserted in a system.

Telephone companies have established frequencies as well as power levels as standards to which their systems are to be tested. North American companies have established a frequency of 1004 Hz with a power level of 0 dBm (1 mW). Most other telephone companies utilize a standard of 800 Hz at 0 dBm. A reading of 0 dBm at the appropriate frequency in either of these systems means that there is no gain or loss. The 1004 Hz 1 mW reference level is called the *transmitted level point* (TLP) in North America, and the 800 Hz 1 mW reference level is called the *decibel reference* (dBr). Table 11-5 provides sample input/output power ratios and decibel conversions.

A sample test problem is illustrated in Figure 11-26. A generator places a 1004 Hz 1 mW signal on a line that has a 2-dB loss.

Table 11-5 A sample representative of decibel losses compared to power losses in electrical circuits

Decibel Values	Power Ratio (Input/Output)
−1 dB	0.79
−2 dB	0.63
−3 dB	0.50
−4 dB	0.398
−5 dB	0.316
−6 dB	0.251
−7 dB	0.20
−8 dB	0.158
−9 dB	0.126
−10 dB	0.100

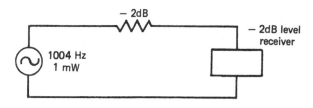

Figure 11-26 TLP level test.

11.15 ZERO TRANSMISSION LEVEL

A system may experience power loss by way of reduced amplification, faulty cable connections, or, in the case of light transmission, a poorly prepared fiber cable connection. To establish the level of reduction in a system as compared with a perfectly functioning system, the term *zero transmission level* (dBm0) is used:

$$\text{Zero transmission level} = \text{actual level} - \text{test level}$$
$$\text{dBm0} = \text{dBm} - \text{TLP}$$

The dbM0 level shows the actual loss in a system as compared with the original specification level. For example, suppose a telephone block has a TLP of 12 dBm but the actual measurement is 8 dBm. Then

$$\begin{aligned}
\text{dBm0} &= \text{dBm} - \text{TLP} \\
&= 8 - 12 \\
&= -4 \text{ dB}
\end{aligned}$$

The system is 4 dB below standard specifications. Using the decibel power formula, the level of power reduction can be determined.

$$\begin{aligned}
\text{dB} &= 10\log P_{\text{out}}/P_{\text{in}} \\
&= -4/10 = \log P_{\text{out}}/P_{\text{in}} \\
&= 0.399
\end{aligned}$$

The power level has been reduced to approximately 0.4 of the power specifications.

Table 11-6 is an example of TLP, dBm, and dBm0 measurement in a system.

Table 11-6 Measuring the TPL of a system

dB/dBm	Output/Input Ratio
−4	−0.399 : 1
−3	−0.523 : 1
−2	−0.699 : 1
−1	−1 : 1
0	1 : 1
1	1.2 : 1
2	1.6 : 1
3	2 : 1
4	2.5 : 1
5	3.2 : 1
6	4 : 1
7	5 : 1
8	6.4 : 1

Table 11-6   Measuring the TPL of a system (Continued)

9	8 : 1
10	10 : 1
16	40 : 1
20	100 : 1
30	1000 : 1
40	10,000 : 1

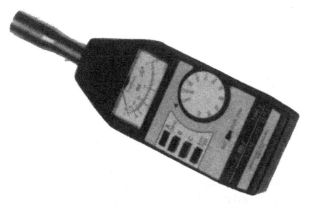

Figure 11-27   Sound/noise measuring system. (*Source:* Courtesy of Simpson Electric Co.)

11.16 *MEASURING INTERNAL SOUND LEVEL*

Excessive noise in the working environment reduces productivity and can be detrimental to workers' health. The maximum noise level for each type of working environment is mandated by the Occupational Safety and Health Administration (OSHA). Sound can be measured by a sound level meter such as shown in Figure 11-27. The instrument can be used to make measurement on any of the three American National Standards Institute (ANSI) weighting curves shown in Figure 11-28. Curve A closely corresponds to the response of the ear and is used by OSHA. Curve B is essentially a flat frequency response that can be used in conjunction with a fast response for an indication of an impulse noise level. Curve C is used to monitor low-frequency noise that needs to be louder to be heard.

11.17 *CATEGORY 5 CABLE TESTING*

In September 1995, TSB[*]-67 was approved to provide a consistent standard for cable testing. The test is designed to validate that a Category 5 link will support applications designed to oper-

---

* Telecommunication Standards Board.

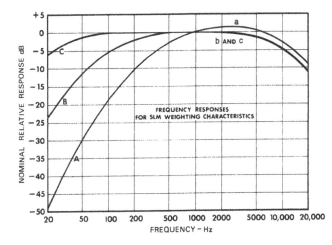

Figure 11-28 ANSI sound weighting curves: (a) closely correspond to the response of the ear, (b) used for impulse noise, (c) used to monitor low-frequency noise.

ate on a generic cabling system. The standard is targeted toward field organizations involved in telecommunication cabling.

TSB-67 specifies that primary field test parameters are as follows:

- Wire map
- Wire length
- Attenuation (see Table 11-7)
- Near-end cross talk (NEXT), far-end cross talk (FEXT), and equal level far-end cross talk (ELFEXT) losses at both ends of the link (see Table 11-8)

The standard specifies two levels for testing. Level 1 is geared toward troubleshooting and occasional testing by administrators. Level 2 is more rigorous testing by installers, contractors, and technicians.

An example of a channel link configuration and test parameters is given in Figure 11-29. The system is assumed to contain a telecommunication outlet/connector, a transmission point, 90 m of UTP wiring, a cross connector consisting of two panels or blocks, and a total of 10 m of patch cable.

11.18 THE TECHNICAL SUPPORT CENTER

A technical support center can help the user via a checklist to determine if the problem involves the use, the user's application, the data terminal equipment, the data communication require-ment, or the link between these applications and the host. Within this process the software is checked to see if proper communications are established between the terminal and controller to host.

To improve performance of 1000Base-T networks, the TIA has proposed Enhanced Cate-gory 5e specifications, as shown in Table 11-9.

Table 11-7   Basic/channel link attenuation

Frequency (MHz)	Category 3 (dB)	Category 4 (dB)	Category 5 (dB)
1	3.2/4.2	2.2/2.6	2.1/2.5
4	6.1/7.3	4.3/4.8	4/4.5
8	8.8/10.2	6/6.7	5.7/6.3
10	10/11.5	6.8/7.5	6.3/7
16	13.2/13.9	8.8/9/9	8.2/9.2
20		9.9/11	9.2/10.3
25			10.3/11.4
31.25			11.5/12.8
62.5			16.7/18.5
100			21.6/24

Table 11-8   NEXT loss (pair to pair)

Frequency (MHz)	Category 3 (dB)	Category 4 (dB)	Category 5 (dB)
1	40.1/30.9	54.7/53.3	60/60
4	30.7/29.3	45.1/43.3	51.8/50.6
8	25.9/24.3	40.2/38.2	47.1/45.6
10	24.3/19.3	38.6/36.6	45.5/44
16	21/19.3	35.3/33.1	42.3/40.6
20		33.7/31.4	40.7/39
25			39.1/37.4
31.25			37.6/35.7
62.5			32.7/30.6
100			29.3/27.1

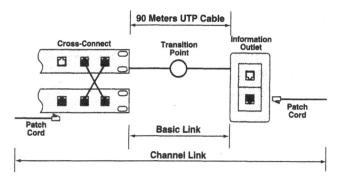

Figure 11-29   Channel link.

Table 11-9   TIA proposed Enhanced Category 5e specifications

Transmission Parameters	Channel Performance
NEXT @100 MHz	27.1 dB
PSNEXT 100 MHz	NA
Attenuation @ 100 Hz	24.0 dB
EFLEXT @ 100 MHz	17.0 dB
PSEFEXT @ 100 MHz	14.20 dB
Return Loss @ 100 MHz	8.0 dB
Propagation Delay @ 10 MHz	555 ns
Delay Skew	50 ns

Once the problem has been isolated, if the tech support center cannot fix it, then it will be referred to a higher level of support. When the problem escalation procedure is initiated the analyst at the tech support center will continue to update and monitor problem status. All technical support personnel involved with the problem should continue to update the problem file. Every entry should be time stamped and initialed by personnel. This provides a history of how the problem was brought to a solution. The end user and management are updated on problem status.

As stated earlier, the information obtained on a problem will serve as a data bank for similar or related problems. Ultimately the information could serve as a diagnostic utility file on a common service system so that other individuals can research a known set of indicators for a possible solution.

The help desk of the tech support center will schedule work, open work orders, track any outstanding problems, review change activity logs for possible service interruptions and/or user service level impact, and issue timely reports for network management. The help desk is typically an administrative and managerial type of job. However, technical requirements can be incorporated into the job description to allow more efficient and competent service at the user's first contact point.

11.19   SUMMARY

The communication technician must be knowledgeable of the range of tests and test instruments for testing ISO protocol levels 1 through 4. However, most functions of a network ride on the fundamentals—cables and connectors.

QUESTIONS

1. What are the simplest instruments that can be used for continuity testing of a cable?

2. What is the purpose of a ground test to a cable?

3. What is the difference between the testing of a coaxial cable and a twisted-pair cable?

4. What special problems are inherent in testing a fiber-optic cable?

5. How is a signal-to-noise ratio measurement performed on a copper cable?

6. Describe your ideal technical support center.

Documenting the Wiring System

12.1 INTRODUCTION

Documentation of the cabling system is paramount for effective service to the user community and maintenance of the telecommunication system. Documentation should consist of all the information and data that would be necessary for you to assume the job of telecommunication manager at a new company from which all the technical staff had left. In other words, documentation must be current, reflecting what is reality at this point in time, and show a history of past events. Current data are necessary for immediate service to the user community and projection of cable availability. Historical data are necessary to assist in management projections of future data processing needs and as a reference for maintenance and troubleshooting by technical staff.

Documentation could be almost solely in the form of a database system. However, backup should be provided in the form of physical diagrams, charts, blueprints, and forms. Most important to any documentation system is the labeling of all the conduits, feeder cables, terminal blocks, circuits, between-building runs, wiring panels, access panels, and so on.

12.2 LABELING THE CABLING SYSTEM

Cabling standards and procedures should be documented and adhered to for all wiring applications. Cable labels should include such items as

- Building number
- Room identifier
- Room cable number
- Media type

Cable label: 011-1-A26-C93

```
XXX  X - XXX  XXX  XXX
1    1    1    1    1
1    1    1    1    1 ----------------------MEDIA TYPE
1    1    1    1                           B = Broadband
1    1    1    1                           Bk = Backbone
1    1    1    1                           Cxx = Coax
1    1    1    1                                 93 = 93 ohms
1    1    1    1                                 75 = 75 ohms
1    1    1    1 ----ROOM CABLE                  63 = 63 ohms
1    1    1
1    1    1    ex. 000-999
1    1    1        A01-299        C = CAD/CAM
1    1    1                       E = Ethernet cabling system
1    1    1                       Fm = Fiber, multimode
1    1    1                       Fs = Fiber, single mode
1    1    1 -----ROOM             I = IBM cabling system
1    1                           Iw = Inside wiring
1    1                           L = Local area network
1    1                           M = Multiplexer line
1    1                           TN = Teleco cable
1    1                                 N = 4 = 4 wire
1    1                                 N = 6 = 6 wire
1    1                                 N = 8 = 8 wire
1    1                           Ow = Outside wire
1    1                           P = Public address system
1    1                           Tw = Twin axial
1    1 ---FLOOR                  Tv = Campus/building TV
1
1
1
1-BUILDING NUMBER

     ex. 000-99
     AO1-299
```

Figure 12-1 Cable labeling standard.

Standards help establish procedures and present connectivity in a logical and understandable manner. This assists the user community, technical staff, and management alike by presenting a clear identifier for each application. Figure 12-1 illustrates a cable labeling standard. An example of proper labeling is as follows:

Cable label: 011-1-A26-C93

In this example, the cable is located in Building 011, first floor office, room A26, room coax #2; it has an impedance rating of 93 Ω; and it is a coax-type cable.

12.3 BLUEPRINTS AND DIAGRAMS

Blueprints are drawings of the physical plant showing the location of all offices, labs, equipment closets, underground tunnels, and so forth. Electrical blueprints are drawings of the same plans but usually have less detail. However, electrical prints may have information not located on building blueprints, such as conduit and cable tray locations. Both blueprints and electrical drawings must be kept up to date as spaces are added and deleted from the facility as wiring changes are made to accommodate the user community.

Furniture drawings are made from blueprints but are less useful because individuals have the habit of rearranging their habitats.

An accurate and up-to-date set of blueprints showing all main distribution frames (MDFs), immediate distribution frames (IDFs), conduit numbers, cable numbers, and equipment locations should be kept up to date by the technical staff. This documentation shows a physical view of the facility and may comprise many drawings for a large campus. This documentation should be the responsibility of one person and reflect the information in the distribution logs and the database.

12.4 DISTRIBUTION LOGS

The distribution log should reflect the current status of the cabling system and should be altered immediately after any changes are made to the system. In a large campus facility the distribution log would probably be several files, one for each building or area. An area or building distribution log might be quite simple, as shown in Figure 12-2. Here the cables (probably twisted pair) are numbered 1 through 12 by their distribution number. Each is identified as to its origin IDF, cable number, port number, and name of user. The user's name may be a person, facility, or device (such as a printer).

The simple distribution log may be further detailed by a wiring closet/controller room log such as the one depicted in Figure 12-3. This is one of several worksheets recommended by IBM for the IBM Cabling System.

Figure 12-4 depicts a detailed form that is useful to the technician installing or testing cabling. This drawing depicts the wiring closet and a rack within the closet. The actual location of each terminal block is shown on the drawing. Each attachment to the terminal block is documented by the type of connection, the accessories in the cable run to the work area, the cable length, and the panel to which the cable is run.

12.5 WORK AREA INVENTORY SHEETS

The documentation sheet depicted in Figure 12-2 is used for a work area or work function, such as an office, lab, or service (such as a research lab)—in other words, a rather small group that has commonality. The form identifies the name of each user within the group; the user's telephone number; the building, floor, and room; the cable utilized; the distribution number; the IDF number; and the types of data processing equipment utilized by the individual. One individual name could represent several employees. In a small company the name could represent a function (such as payroll).

12.6 HANDWRITTEN ENTRY VERSUS TERMINAL-BASED ENTRY

Historically organizations have used handwritten entries to update their tracking and inventory systems. Although the setup time was very labor intensive, once in place it worked well pro-

Dist. Circuit No	IDF	Cable No	Ext./Port	Name
1	A			
2	A			
3	A			
4	A			
5	A			
6	A			
7	A			
8	A			
9	A			
10	B			
11	B			
12	B			
13	B			
14	B			
15	B			
16	B			
17	B			
18	B			
19	B			
20	B			

Figure 12-2   Sample distribution log of a small building or a work area within a building.

vided that the facility or inventory didn't change often. However, once change began increasing in frequency, the handwritten entry revealed several defects.

The most common among these defects is the problem of maintaining current and accurate records during times of high activity. With moves, adds, and change occurrences, this becomes a large task. Most organizations maintain a master log and distribute copies of this log to the various work or activity centers. Usually, tracking updates are placed on a department calendar for weekly or monthly reviews and audit procedures are performed weekly or monthly.

This system works well when things are rather static and running smoothly. However, when the workload is high, installers or technicians may not have the time to make log entries. The notes of the changes may become a wad of paper in someone's pocket. Once this happens, management will usually assign the update task to one person. This means increasing head count and adding one more step to the process, and therefore one more potential failure point.

When the system begins to break down, other tracking and updating alternatives will be proposed. Among them will be the cable database management system. This system works well for updating and maintaining files as well as providing current data to all personnel who have access authorization. Chapter 13 discusses the components of such a database.

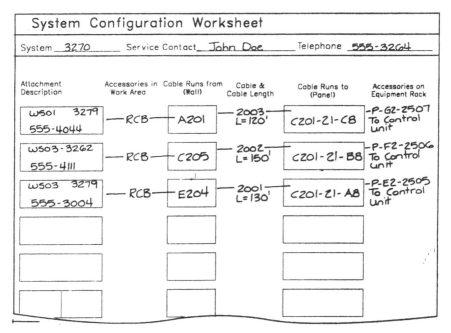

Figure 12-3 Wiring closet/controller room worksheet. (*Source:* Courtesy of IBM, Inc.)

12.7 SUMMARY

The examples discussed in this chapter are intended to serve only as ideas for evaluating your organization's current documentation or as a starting place for new documentation. The amount of detail necessary for documentation could be debated for the remainder of the text. However, each telecommunication manager must decide what is right for his or her company by considering factors such as age of the existing system, number of cables, distribution systems, other hardware, and the requirements of upper management. Any documentation should also be

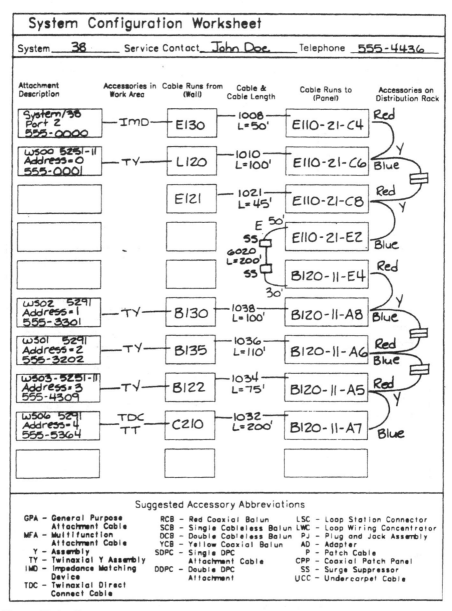

Figure 12-4   System configuration worksheet. (Source: Courtesy of IBM, Inc.)

designed to meet the needs of the wiring system database, discussed in Chapter 13. There must be sufficient documentation to allow the technical staff to function in a timely manner and for management to make expedient economical decisions.

QUESTIONS

1. What necessitates the establishment of a wiring documentation system?

2. When is a written file documentation system justified?

3. Who should be responsible for a file-based documentation system?

4. What factors could make a file-based documentation break down?

CHAPTER 13

Telecommunication Database

13.1 INTRODUCTION

The purpose of this chapter is to outline the requirements of a telecommunication-specific database and to give an example of one such system.

Every wiring or cabling system for a telecommunication network requires some form of tracking facility. To be effective, tracking must be capable of update functions (for adding, modifying, and/or deleting records) and retrieval functions (based on request inquiries). The system must also be able to generate inventory and report responses. The "tracking tool" must be capable of allowing the **management information systems** (MIS) departments and telecommunication technical support personnel access to the information. Each group or function will have different reasons for access. Therefore, there should be different access levels controlled by a central person:

- The technical support group will need to be able to do a search based only on fields. The only information they may have might be the room information or the user's name. The system will have to provide the rest of the information, such as cable and host connection data.
- Management concerns may include the total connectivity throughout the loading and host channel utilization.
- The control center will have to be able to answer inquiries from users, management, and IS organizations based on very little data concerning host service connections, types of cable, and possibly overall lengths of specific wire runs.

There are basically two types of tracking tools to perform the aforementioned functions: file-based systems and database systems. These two systems are discussed in the following sections.

13.2 FILE-BASED TRACKING SYSTEM

Traditionally, file-based systems have been used for tracking because of their simplicity and the fact that LANs have been rather small. When the LAN is in a small building or facility, the various systems and system connections are simple and a file-based system will probably perform effectively. The decision to select file-based or database tracking depends on the size and complexity of the wiring network (existing and future requirements) and the degree of control that management wishes to have over the system. It also depends on the future product offering or host and/or system requirements. In the case of a small user community that is fairly stable, the file-based tracking method could satisfy both users and the information system department.

As elaborate as file-based systems can become, they all share certain inadequacies for a large, complex facility with many users. Since there is no common data file, there will be many "copies" of common files. This redundancy carries with it the obvious concern for updating and/or revising any one file, which means that all copies must be updated. A second concern is with regards to commonality as it is important that all changes carry identical definitions and common factors. Otherwise, a correct update to one file may not be accurately reflected when another file is revised. Also, because many application programs will be accessing various files, there will be some programs with logic overlap. Even with the use of sort/merge programs, time and resources will be expended in different areas to accomplish the same end task.

If the current system cannot keep pace with the growing number of changes and updates or the host service system is being updated on a regular basis to handle the increased user load, then the file-based system may no longer be adequate to meet the requirements.

13.3 DATABASE MANAGEMENT SYSTEM

The **database management system** (DBMS) provides for integrated accessing, linking (implicit and explicit), and managing of data files. The database management system provides for a user-friendly interface for nonprogrammers to do inquiries and retrievals from shared or common files.

There are three types of DBMS organizations: hierarchical, network, and relational. For our purposes, the most versatile type will probably be the relational database model. In this model all data are located and written in tables. These tables improve flexibility by organizing data in small, related groups. When a query comes into the database the table is looked up and data are presented. In the case of a large or complex request, small tables are organized into one large table to meet the needs of the request.

The reason for the small table size format is that larger tables would necessarily have to contain data that would be duplicated elsewhere in other large tables. This duplication of data would add to the overload of the process and would not be an efficient data storage structure.

The DBMS is very flexible and can handle requests that it may not have been designed to address due in part to the use of small data tables and implicit and explicit links. The management system will also provide a certain amount of isolation between the data files themselves and the user community through the inquiry language.

13.4 STRUCTURED QUERY LANGUAGE

In 1986, ANSI defined and standardized the language requirements for relational databases by specifying that **Structured Query Language** (SQL) would be the standard language for all relational database management systems.

SQL is a query language that can be used to create and maintain the database structure. It will also allow data manipulation. SQL allows the user community to specify the data field or group of data requested; the management system will take the inquiry and search for the requested data from the data tables. Once enough data have been assembled to meet the request, the links and the data are provided. Depending on how the database is structured and oriented, the requested data (which may be only one data field) can be obtained from one or more inquiry panels or menus.

The data tool gives the telecommunications personnel the ability to log onto an online system and perform a number of search inquiries.

Probably two of the most significant uses for such a database are current connectivity status and forecasting. Connectivity status informs the telecommunication personnel as to what is connected to the system, how it is connected, and to whom it is assigned. The forecasting capability allows the TCM to know exactly the number of cables in the system, their current usage, and future capacity.

13.5 BASIC COMPONENTS OF A DATABASE MANAGEMENT SYSTEM

There are three basic components to the database management requirement for a telecommunication system:

1. Resources required to plan, write, and maintain the database
2. A database administrator (DBA)
3. A contact point for all telecommunication requests for cable or wire connection to host services

Although the initial cost of creating a database management system is greater than that of a file system, its strengths far outweigh the additional expense considerations. Some of the advantages of the DBM system over the file system are as follows:

1. The database offers greater access to organized data than the file system through the use of host and query languages.
2. Data can be accessed from either application programs or query languages that can be written for the nonprogrammer. This ease of use helps all concerned to obtain the data or records needed without the requirement of advanced language skills.
3. The user can log onto an interface that will simplify inquiries on the database for the retrieval or update of information.

4. The database can be used for technical support as well as planning and management for the proper and timely coordination of premise wiring systems.

5. Not only will the database provide an accurate picture of current usage, but it will also provide information on the current product mix of PCs, workstations, printers, and terminal use.

13.6 DATABASE MANAGER'S RESPONSIBILITY

The database manager (DBM) is responsible for the database. He or she must ensure the accuracy of the data for tracking, inventory, and forecasting purposes. The DBM may have been involved with the writing of the program and/or the organization of the files or records that built the system. In either case the DBM must ensure the effectiveness and efficiency of the data retrieval system.

The DBM has the authority to assign read-only or read-and-write access to the database. Management and technical support require different levels of access to the database depending on their job assignments. The DBM is responsible for setting minimum performance measurements, including system response time.

The DBM may have been or may become involved in writing the database program or program requirements. Regardless, the DBM must ensure that:

1. The system is designed with enough flexibility to evolve and accommodate upgrading with a minimal impact on service levels.

2. A contact point is maintained by the technical support center to account for all requests for new service and/or cable runs.

3. The database is used by both technical support as well as planning and management for the proper and timely coordination of premise wiring systems.

4. The database provides an accurate picture of current product mix of PCs, workstations, and terminals in use.

13.7 SAMPLE DATABASE

To show how a sample database can be constructed and what it will look like to the user community, technical support center, and management, some basic assumptions about the design and layout are given in this section.

This example is a sample only. In the real world there would be several meetings between affected and interested groups to determine the database design, how it would be laid out, who would have access, and myriad other individual and corporate concerns, special requirements, and needs.

The first issue is to determine if the system will be *user data driven* or *host service driven*. A user data driven facility would accommodate mixed hardware, such as Apple, IBM, HP, and/ or DEC workstations all connected to the network. In other words, the user controls the hardware and the host then sets up the protocol and proper connections. In a host service driven facility, the hardware and software are predetermined for the community. Any workstation or terminal hookup is specified to a certain group of hardware. All users coming into the facility would use IBM, DEC, HP, or Apple workstations. The host dictates the protocol. There would be no opportunity to connect mixed vendor PCs or workstations on the existing network.

Our database example will be menu or screen driven. It will consist of five major panels to be used for inquiry, data retrieval, updating, and initiating a work request.

- First screen: *Master inquiry panel*
- Second screen: *Location/room cable panel*
- Third screen: *Link and connect panel*
- Fourth screen: *Service availability panel*
- Fifth screen: *Work order panel*

The master inquiry panel (Figure 13-1) is a read-only file, although all fields are searchable within the panel. The original data for this panel should have been obtained from verified records or by a physical inventory of the existing facility and an accurate documentation of all cable connections from the host to the end user's room termination by the department in charge. This allows various organizations to have access to this panel for information and status update or related activities within their functions. Users within selected organizations can directly access this panel and have restricted write access so that they can submit requests online. The panel can also be used for the following purposes:

- By support center personnel to schedule work, track a request for service or a device, and initiate host or remote system connections
- To monitor current connectivity loads by system. This will help in anticipating bottlenecks and possible performance problems.
- To support center personnel to schedule work, track a request for service or a device, and initiate a host or remote system connection request.
- For management to inquire about the availability of existing cable runs within a given department or number of offices. This may be for resource consolidation (double up personnel in offices) or to upgrade service levels in a particular group.
- For management approval when an employee needs access to a secure system or increased service support levels that need prior management approval.
- For special requirements. For example, suppose there is a litigation department that is dealing with leased lines for specific Japanese products. These product offerings will be designated for the overseas market and all application programs will be written for Japanese-speaking clients. The development computers will need to run under Japanese language requirements. Because of this, a small number of controllers will be specifically configured and adapted for that purpose. However, to set up an application system both the controller and the workstation have to be specified or the program will not run properly. In

---

MASTER INQUIRY PANEL

USER DATA

1) USER NAME: (LAST) _____ (FIRST) _____
   USER'S OFFICE PH: _____ COMPANY I.D. _____
2) LOCATION: (BLDG) _____
   ROOM NUMBER: _____
3) DEPT NUMBER: _____ DEPT MANAGER: _____
   DEPT MANAGER PH: _____

ROOM CABLE DATA

4) NUMBER OF CABLES IN ROOM:
   ( ) 1,   ( ) 2,   ( ) 3,   ( ) 4,   OTHER ( ): CABLE # _____
   NUMBER OF CABLES UNUSED IN ROOM:
   ( ) 1,   ( ) 2,   ( ) 3,   ( ) 4,   OTHER ( )
5) TYPE OF ROOM CABLE AVAILABLE:
   CABLE #1: TWISTED PR ( ),   COAX ( ),   FIBER ( ),   LAN ( )
   CABLE #2: TWISTED PR ( ),   COAX ( ),   FIBER ( ),   LAN ( )
   CABLE #3: TWISTED PR ( ),   COAX ( ),   FIBER ( ),   LAN ( )
   CABLE #4: TWISTED PR ( ),   COAX ( ),   FIBER ( ),   LAN ( )
6) ROOM CABLE #1 IDENTIFIER: _____
   ROOM CABLE #2 IDENTIFIER: _____
   ROOM CABLE #3 IDENTIFIER: _____
   ROOM CABLE #4 IDENTIFIER: _____

TERMINAL/DEVICE DATA

7) NUMBER OF TERMINALS AND/OR DEVICES IN ROOM:
8) TERMINAL/DEVICE TYPE, SERIAL NUMBER (S/N) AND MODEL
   DEVICE #1 TYPE: _____ S/N: _____ MDL: _____
   DEVICE #2 TYPE: _____ S/N: _____ MDL: _____
   DEVICE #3 TYPE: _____ S/N: _____ MDL: _____
   DEVICE #4 TYPE: _____ S/N: _____ MDL: _____

TYPE OF REQUEST SECTION

9) NEW SERVICE: ( )
10) RECONNECT SERVICE: ( )
11) DELETE EXISTING SERVICE: ( )

Figure 13-1 Master inquiry panel.

WORK ACTIVITY SECTION

12) INSTALL: ( )
13) EQUIPMENT NEEDED: _____
    _____

14) ESTIMATED TIME OF ARRIVAL (ETA): _____
15) REMOVAL: ( )

TYPE OF CONNECTION REQUESTED

16) DIRECT CONNECT (HARDWIRED): ( )
17) TERMINAL CONCENTRATOR: ( )   IF CHECKED CONC. S/N: ___
18) FRONT END PROCESSOR: ( )   IF CHECKED FEP S/N: _____

TYPE OF HOST REQUIRED

19) FLOOR SYSTEM (NATIVE): ( ) _____
20) REMOTE SYSTEM: ( ) _____
21) DEDICATED SYSTEM: ( ) _____
22) ISOLATED SYSTEM: ( ) _____
23) ENGINEERING SYSTEM: ( ) _____
24) TEST SYSTEM: ( ) _____

I.S. INTERNAL USE

25) RESTRICTED ACCESS: YES ( )   NO ( )
26) AUTHORIZING MANAGER: _____   PH: _____
27) CONTACT NAME: _____
28) CONTACT PHONE NUMBER: _____

END OF MASTER MENU

Figure 13-1   Master inquiry panel. (Continued)

this example the requester would specify which device is needed, which control must be connected, and so on.

13.8 FIELD DESCRIPTION FOR MASTER PANELS

The master panel is broken into nine categories:

User data
Terminal/device data

Type of request
Type of connection required
Type of host required
Specific system
Specific controller/port controller
Special needs
IS only

- *Field 1: User's name* This field is reserved for the user's name or contract. It should be entered as the last and then first name. In this format, a search routing or inquiry can easily locate the correct user.
- *Field 2: Location* This field consists of both the *building* (BLD) *number* and *room number*. In a search routine based solely on *room number*, the analyst should be able to obtain the following information:
    1. Number and type of cables existing in room.
    2. Type of host and controller service available on which room cables.
    3. Type of terminals and/or devices already in place in the room. Typically, this is the most often used search to provide information to the requester.
- *Field 3: Department number* This field is really for tracking purposes. For example, if a department was physically located within one building, or perhaps on one floor or core area, all host/controller connections could be located and routed through common controller services. A good example of this would be a litigation department or a project design team. All online resources could be located on a single (or few) dedicated controller(s). This dedication has some immediate advantages of security and commonality to the functional unit.
- *Field 4: Room cable number* This information, once input to the system, will allow for specific searches to be done on a particular cable, coax, or fiber line. On the query screen, the rest of the data will be presented. In this manner, questions such as whether a particular cable is active, what service is available (if any), and so forth can be addressed quickly.
- *Field 5: Room cable type* This field specifies the type of cable room available.
- *Field 6: Cable labeling information* This field provides corporate cabling information.
- *Field 7: Terminal or devices* This field shows the number of devices in a room.
- *Field 8: Terminal or device data* This field is for the numeric identifier for the particular device or terminal in the office that is currently there or is being requested.
    1. *Serial number* This is the serial number of the device or terminal. Fields 10 and 11 are excellent resources to conduct a quick audit of physical assets within a department or function. Accounting departments may want to get a monthly or quarterly report with at least these two fields included.
    2. *Model* This will be the model of the terminal or device in the user's office or laboratory.

13.9 TYPE OF REQUEST SECTION

- *Field 9: New service* This selection is used for new service. When marked with an X, this begins the new service work activity. The analyst and technician will work off this order as well as the room cable panel and connect panel to complete the installation service.
- *Field 10: Reconnect service* This field is used for reconnecting a deleted service.
- *Field 11: Delete existing service* This field is used to delete a service when, say, an employee is vacating an office or lab.
- *Field 12: Install* This is an install order. Once this box is checked, all telecommunication personnel will know that this is an order for a hardware install.
- *Field 13: Equipment needed* This is for hardware needed, such as terminals, printers, and so on.
- *Field 14: Estimated time of arrival (ETA)* This is a data field. It is used when equipment or devices are being sent in from different locations. It can be as specific as the actual arrival date at the user's office or room or it may be only a guess as to when the needed items will arrive.
- *Field 15: Removal* If checked with an X, this request is to remove a piece of equipment or device. The analyst and tech support group will have to set up some form of tracking, inventorying, and storage facilities for this purpose.
- *Field 16: Direct/connect hardwiring* This field is used to identify those devices that require direct connection or hardwiring, usually to a floor or native system.
- *Field 17: Terminal concentrator* This field is used to identify a request for a terminal concentrator connection.
- *Field 18: Front-end process* This field is the main processor.
- *Field 19: Floor native* When marked, this indicates that the user requires a floor system connection.
- *Field 20: Remote system* When marked, this indicates that the user requires a remote system connection.
- *Field 21: Dedicated system* When marked, this indicates that the user requires a dedicated system connection.
- *Field 22: Isolated system* When marked, this indicates that the user requires a connection to a standalone system.
- *Field 23: Engineering system* When marked, this indicates that the user requires a connection to an engineering system.
- *Field 24: Test system* When marked, this indicates that the user requires a connection to a test system.
- *Field 25: Restricted access* There can be a question of security regarding access for telecommunication personnel to the user's office or room. If there is restricted access, this field should be marked yes and Field 26 (contact) should be filled in with the person's name as well as Field 27 (contact phone number) for the phone number of the person to be contacted.

- *Field 26: Contact* This is the person to be contacted for access to the requester's room or office. This field is used when there is restricted access concern or the user is out of the office and this person has a key to an access approval.
- *Field 27: Contact phone number* This is the phone number of the contact.

Specific controller and/or port requirements are other areas of unique information to specify a controller that is configured on a certain microcode level. The port information may be requested for special PCs or workstations that need to be port configured or customized to meet a unique device or terminal requirement.

For facilities that have three or more point network strategies, the panels on the following pages will assist in completing the connection paths and links between user device and host.

To complete this example we must take some liberties with the fictitious facility that is being described. For instance, there is a minimum of one and a maximum of five cables in each office or laboratory space. The location/room cable panel (Figure 13-2) can be used quite effectively when the facility is primarily host service based. The services that are available are indicated within each office or lab space. If enough system hardware is available, a "hot service" concept can be implemented to provide at least one live service connection within each room on a designated room cable for service connection to a user's device.

The major tracking point within this strategy is the room cable. The room cable information is important because it allows any person with read-only access to check on service level availability by room cable data. Some organizations will only track connect points or hosts to room cable links and disregard device hardware. In this instance the primary concern is with controller connections, controller microcode levels, and device support groupings within a defined product range. An appropriate database and support service is developed to maintain the system.

The location/room cable panel (Figure 13-2) is the second screen panel in the database series. The panel contains the following:

- *Field 1: Building number* This format is the same as for the master inquiry panel.
- *Field 2: Floor* This field is a three-digit field for alphanumeric data.
- *Field 3: Room number* This format is the same as for the master inquiry panel.
- *Field 4: Room cable number* This field is a four-digit alphanumeric number. There is a total of five fields to support up to five cable runs within the same office or lab space. For example:

$$\text{cable } 110 = -110$$
$$\text{cable } 02 = -002$$
$$\text{cable } 14/7 = -14/7$$

With the information provided by the building, room, and cable number fields, the technical support personnel can go directly to the cable selection panel for all room cables run into this room and determine which ones are active.

```
| LOCATION/ROOM CABLE PANEL
|
|   1) BLDG NUMBER: _____
|   2) FLOOR: _____
|   3) ROOM NUMBER: _____
|   4) ROOM CABLE NUMBER: _____
|   5) CABLE TYPE: _____
|   6) CONNECTION COUNT: _____
|   7) CONTROL UNIT TYPE: _____
|   8) CONTROL UNIT SERIAL NUMBER: _____
|   9) HOST SYSTEMS: _____
|  10) WORK REQUEST ORDER NUMBER: _____
```

Figure 13-2   Field description for location/room cable panel.

- *Field 5: Cable type* Refer to Chapter 12 (Section 12.2).
- *Field 6: Connection count* This field is optional. When filled in it will give the number of connections between the controller and the end user's office. Another variation of this is the connection count between the host and end user's office.
- *Field 7: Control unit type* This information will assist management and technical support in tracking the connectivity base. It also gives information concerning the type of controller and microcode level to determine the support service level for the device or product, LAN or WAN, and so forth.
- *Field 8: Control unit serial number* (if applicable) The serial number of the controller can be input to specify a controller.
- *Field 9: Host system* This can be information concerning a terminal concentrator or a direct connect, dedicated system.
- *Field 10: Work request order number* If a work order is indicated, then a request may be outstanding for a hookup or service connection.

The room cable panel is the list of all cable numbers assigned but not always present in the room. It is important to note that cable room numbers can start at any point since we have designated that all room cables carry a standardized cable number and type of cable (Chapter 12).

13.10 FIELD DESCRIPTION FOR LINK AND CONNECT PANEL

The third screen in the series (Figure 13-3) contains the following:

- *Field 1: Building number* This format is the same as for the master inquiry panel.
- *Field 2: Room number* This format is the same as for the master inquiry panel.
- *Field 3: Room cable number* This format is the same as for the master inquiry panel.
- *Field 4: Host system* This will be the actual dedicated host system or terminal concentrator.

```
┌─────────────────────────────────────────────────────────────┐
│ LINK AND CONNECT PANEL                                        │
│                                                               │
│  1) BLDG NUMBER: _____            │
│  2) ROOM NUMBER: _____            │
│  3) ROOM CABLE NUMBER: _____            │
│  4) HOST SYSTEM: _____            │
│  5) CHANNEL FIELD: _____            │
│  6) CONTROL UNIT TYPE: _____            │
│  7) CONTROLLER SERIAL NUMBER: _____            │
│     MICROCODE LEVEL: _____            │
│  8) CONTROLLER FLOOR LOCATION: _____            │
│  9) CONTROLLER PORT ADDRESS: _____            │
│ 10) PORT CONFIGURATION: _____            │
│ 11) CONTROLLER BREAKOUT PANEL RACK NUMBER: _____            │
│ 12) RACK AND FIELD POSITION: _____            │
│ 13) PANEL TERMINATION: _____            │
│ 14) COMPUTER FLOOR TERMINATION: _____            │
│ 15) REMOTE ROOM PANEL: _____            │
│ 16) WORK ORDER NUMBER: _____            │
└─────────────────────────────────────────────────────────────┘
```

Figure 13-3   Link and connect panel.

- *Field 5: Channel field* This field is optional. When filled in it indicates the number of the channel to which the controller is attached.
- *Field 6: Control unit type* There are various controller types each with specific characteristics.
- *Field 7: Controller serial number* This is the controller's serial number.
- *Field 8: Controller floor location* This is a six-character field for alphanumeric data describing the physical location of the controller.
- *Field 9: Port address* This is the actual port address as defined by the system.
- *Field 10: Port configuration* This is the port definition for this specific port.
- *Field 11: Controller breakout panel rack number* This is the location of the panel where a 32-port controller is broken out into 32 individual port terminations.
- *Field 12: Rack/field position* This is the location of Field 11, if the panels were put into a subrack system.
- *Field 13: Panel termination* This is the actual port terminal on the rack or field panel.
- *Field 14: Computer floor terminal* This is the location of the computer cable terminations. In a rack or field mounted system, this is the host panel.
- *Field 15: Remote room panel* This is the location of the panel where the remote room cables terminate. In this case the remote panels are not located at or in the computer and controller floor area but somewhere else in the facility.

• *Field 16: Work order* If filled in, this is the work order associated with this cable run.

13.11 SERVICE AVAILABILITY PANEL

The service availability panel is used for data retrieval. The cable data will be entered in the request portion of the panel. This information will be known from the location/room cable panel (Figure 13-4).

In response to the request, the database system will provide the reply data. This information will include the controller type, its serial number, the controller microcode level, and the system host that is on the specified cable number. The other portions of this panel are repetitive for those rooms or labs with more than one cable. If there is not a host service connection on a particular cable, that will also be noted.

```
SERVICE AVAILABILITY PANEL

    REQUEST

#1) CABLE NUMBER: _____
                                    REPLY TO REQUEST #1
                                    CTLR: _____    MODEL: _____
                                    S/N: _____
                                    MICROCODE LEVEL: _____
                                    LOCATE COORDINATES: _____
                                    TOTAL # OF CNTLR PORTS: _____
                                    TOTAL # OF AVAIL PORTS: _____
#2) CABLE NUMBER: _____
                                    REPLY TO REQUEST #2
                                    CTLR: _____    MODEL: _____
                                    S/N: _____
                                    MICROCODE LEVEL: _____
                                    LOCATE COORDINATES: _____
                                    TOTAL # OF CNTLR PORTS: _____
                                    TOTAL # OF AVAIL PORTS: _____
#3) CABLE NUMBER: _____
                                    REPLY TO REQUEST #3
                                    CTLR: _____    MODEL: _____
                                    S/N: _____
                                    MICROCODE LEVEL: _____
                                    LOCATE COORDINATES: _____
                                    TOTAL # OF CNTLR PORTS: _____
                                    TOTAL # OF AVAIL PORTS: _____
           END OF SERVICE AVAILABILITY PANEL
```

Figure 13-4 Service and availability panel.

The service availability panel will display up to four cable-to-host connections in our example. The program could be written for as many cables as required. This panel is very useful to both support personnel and the user community because each can look up the current status of host services by the cable number.

13.12 FIELD DESCRIPTIONS FOR THE WORK ORDER PANEL

The fifth panel is the work order panel (Figure 13-5). This panel is based on the master inquiry panel except that this is a read/write panel. Although it has much of the same information dupli-

---

WORK ORDER PANEL

ORDER #: _____

USER DATA

1) USER NAME: (LAST) _____   (FIRST) _____
2) LOCATION: (BLDG) _____
3) ROOM NUMBER: _____
4) ROOM CABLE NUMBER: _____
5) DEPARTMENT NUMBER: _____
6) OFFICE PHONE NUMBER: _____
7) DEPT NUMBER: _____   DEPT MANAGER: _____
                              DEPT MANAGER PH: _____

ROOM CABLE DATA

8) NUMBER OF CABLES IN ROOM:
   ( ) 1, ( ) 2, ( ) 3, ( ) 4, OTHER ( ): CABLE # _____
   NUMBER OF CABLES UNUSED IN ROOM:
   ( ) 1, ( ) 2, ( ) 3, ( ) 4, OTHER ( )
9) TYPE OF ROOM CABLE AVAILABLE:
   CABLE #1: TWISTED PR ( ), COAX ( ), FIBER ( ), LAN ( )
   CABLE #2: TWISTED PR ( ), COAX ( ), FIBER ( ), LAN ( )
10) ROOM CABLE #1 IDENTIFIER: _____
    ROOM CABLE #2 IDENTIFIER: _____

TERMINAL/DEVICE DATA

11) NUMBER OF TERMINALS AND/OR DEVICES IN ROOM:
12) TERMINAL/DEVICE TYPE, SERIAL NUMBER (S/N) AND MODEL
    DEVICE #1 TYPE: _____   S/N: _____   MDL: _____
    DEVICE #2 TYPE: _____   S/N: _____   MDL: _____

Figure 13-5 Work order panel. (Continued on facing page.)

TYPE OF REQUEST SECTION

13) NEW SERVICE: ( )
14) RECONNECT SERVICE: ( )
15) DELETE EXISTING SERVICE: ( )
16) INSTALL: ( )
17) EQUIPMENT NEEDED: _____

18) ESTIMATED TIME OF ARRIVAL (ETA): _____
19) REMOVAL: ( )

TYPE OF CONNECTION REQUESTED

20) DIRECT CONNECT (HARDWIRED): ( )
21) TERMINAL CONCENTRATOR: ( )   IF CHECKED CONC. S/N: ____
22) FRONT END PROCESSOR: ( )   IF CHECKED FEP S/N: _____

TYPE OF HOST REQUIRED

23) FLOOR SYSTEM (NATIVE): ( ) _____
24) REMOTE SYSTEM: ( ) _____
25) DEDICATED SYSTEM: ( ) _____
26) ISOLATED SYSTEM: ( ) _____
27) ENGINEERING SYSTEM: ( ) _____
28) TEST SYSTEM: ( ) _____

I.S. INTERNAL USE

29) RESTRICTED ACCESS: YES ( )   NO ( )
30) AUTHORIZING MANAGER: _____ PH: _____
31) CONTACT NAME: _____
32) CONTACT PHONE NUMBER: _____

SPECIFIC SYSTEMS, CONTROLLER, PORT OR CONFIG REQUIREMENTS

_____
_____

WORK ORDER DATA

WORK ORDER NUMBER: _____ DATE ASSIGNED: _____
RESPONSIBLE DEPT: _____ ESTIMATED TURNAROUND DATE: _____

cated on it from the master inquiry panel, it differs in that the analyst can, through the hand-shaking routine of other data tables, assign controllers, specify host systems, and choose from other options on the panel.

This information panel is filled out by the analyst who runs the support center. Once all the required data fields are filled in, the system will assign a work order number. This number will be used to track the work to be performed with time-stamped entries from the various individuals or groups to which the work is assigned.

13.13 SUMMARY

The development of an effective, complete database system detailing cabling, data equipment type and location, user location and identification, maintenance, system utilization, and growth projection is necessary for efficient operation of a telecommunication system. Such a database can help to smooth out the daily problems that occur in any telecommunication system. The database will allow for better utilization of existing cabling, equipment, and time. In doing so, not only can it improve harmony among employees, but it can also increase profit.

QUESTIONS

1. What is the purpose of a wiring database?
2. Identify the minimum functions that you would require in a database.
3. Compare host-driven and user-driven systems.
4. Develop a wiring distribution table for the LAN in Figure 9-23.

CHAPTER 14

Managing Wiring Problems

14.1 INTRODUCTION

The telecommunication manager is faced with many problems, including the following:

1. Evaluating existing voice and data networks to ensure that they meet current and future needs
2. Dealing with office moves, adds, and changes
3. Growing network and hardware (PC, workstation, etc.) requirements
4. Changing network services (expansions and upgrades)
5. Handling company growth and resource consolidations
6. Staying current with new product offerings and evaluating them in terms of cost/benefit
7. Maintaining network security
8. Making an economical business decision for new equipment and products or the upgrade of current equipment
9. Presenting decisions to upper management
10. Introducing new technologies and products into the connectivity network with the lowest possible integration impact

To compound these problems, the end user's requirements (imagined and real) are a moving target for the telecommunication manager. The current inventory of user products and devices can include major brand PCs (IBM, HP, DEC, and so on), PC clones (with varying degrees of compatibility), workstations, specialized printers, multifunction phones, multiplexers, plotters, multiple host service or special dedicated host services, LANs, Ethernets, dumb terminals, intelligent terminals, CAD and high-level graphics terminals/PCs, and specialized language devices. The list could go on. Each of these devices has specific and often different wiring and cabling requirements.

Historically, the premise cabling system has gone unrecognized as a separate resource. Typically, the cabling requirements and the associated costs were developed as a part of, or a requirement for, another product or device. The separate cabling system was rarely considered as a product itself.

The wiring and cable system for premise and off-premise interconnect networks has been overlooked by management. Thus management and technical functions lost sight of the reuse value of the existing interconnect network. Corporations have failed to identify the significance of currently installed wiring systems. Existing cabling networks can make a contribution through many product development cycles.

This oversight is not surprising because the areas of responsibility are often ill defined for the telecommunication manager. Managers have a three-part problem:

1. The business rationale for specific hardware and software products must be determined. Business justifications must be put together and complemented by an overall plan toward greater utilization efficiencies of current resources. In view of the current state of technology, there is more than a good chance that any product purchase will be outdated prior to its life cycle and therefore an argument could be made for staying out of current marketing and waiting. This argument can always be made about any technically based product. However, with good managerial direction based on a clear, no-nonsense approach to telecommunications networks, this argument can be nullified.
2. Once all equipment has been ordered and installed, the impact on the existing system must be understood in terms of system performance versus service interruption. In many data processing shops any service-level deterioration cannot be tolerated unless in emergencies. Management must be able to qualify both service interconnection time (and place this time requirement during off-prime shift periods) and value-added function or service.
3. All product installation, including interconnect networks, must be inventoried and tracked as part of an ongoing process to document the data processing facility. Once the documentation is up to date and current, product and floor planners can take better advantage of all systems and networks. Both components are resources that can be upgraded to meet future requirements, but without proper documentation and measurement points, the resources at hand cannot be utilized fully.

14.2 TRACKING

Current wire and cable systems need to be documented and tracked (Chapter 12). This should be done with proper documentation forms and on a database program (Chapters 12 and 13).

14.3 MEASURING, TESTING, AND TROUBLESHOOTING

When a wiring system or any part of a wiring system is changed, it must be tested by the appropriate technical personnel (Chapter 11). Once the interconnect networks are tested and documented, measurement points can be established and monitored to determine system loads, application factors, and areas of underutilization.

14.4 RETROFITTING

When retrofitting or upgrading a facility, the concept of "highest possible telecommunication application" should be discussed in all meetings and discussion groups (that is, the concept of maximum utilization of current or projected facilities, whether it is a small office space or a large laboratory building).

14.5 COST FACTORS

Cost factors should be balanced against current facilities utilization and possible future applications. In this manner the additional cost for more functions can be determined in advance versus the estimated cost of upgrade retrofit at some future date.

14.6 DATABASE DEVELOPMENT

Some thought must also be given to database development. It must be determined if the facility is to be user data driven or host service driven. The answer will have far-reaching effects on the level of service provided and the type of database that is written.

Questions that must be answered to determine the proper type and level of database are as follows:

- Will the IS and telecommunications departments track only cable runs and the cable host service levels irrespective of the user community or will the responsible departments decide to control all data fields, connection points, and devices with the network?
- What part of the interconnect network will the telecommunications department have control over? What part will the IS department control?
- Who will be responsible for performing overall quality checks and standards for system performance interconnect network control, maintenance, and supervision?
- Who will manage the cable networks, including premise and off-premise networks?
- Who will install, update, and maintain LANs and all new telecommunication technology?
- What level of service is to be provided and in what areas?
- What provider service and turnaround time is to be allowed?

Obviously, any telecommunications tool will be able to track all data fields, but it is far more complex for a single department to control and maintain user devices, user connections, host service access levels, and wire and cable management. The answer to these questions will set in motion the structure of the telecommunications department responsibilities and the requirements for the entire user community.

14.7 NETWORK SECURITY

Network security becomes increasingly difficult as everyone—suppliers, customers, consultants, and employees—increasingly needs to access data stored in the mainframe, often from outside the local network firewall. Managers and communication specialists need to rethink the concept of security from a local control problem to a global control problem. The present techniques of individual passwords are not working, partly because of human nature, in the handling of pass-words. The problem needs to be taken out of the realm of human nature and into technology.

Computer-to-computer security is becoming increasingly secure with an industry security technology called the secure socket layer (SSL). However, human-to-computer access can be conducted by unauthorized and undetected intruders because unique user authorization is not ensured by SSL.

A solution by Identix, Inc., called *Authenticated Certificate*, promises a solution by using the unique characteristics of each person's fingerprint as the identifier. The fingerprint is an identifier that is unique to each person, can only be in the individual's possession, and is always available. Such an identifier would provide security to credit card holders, banks, and Internet merchants.

Trials with communities of a few thousand have proven successful. With the Identix sys-tem, the unique fingerprint characteristics of a user, called *minutia*, are converted to a com-pressed 128-bit data stream and sent to the mainframe for validation. The problem remains of embedding the fingerprint characteristics of every user into the system and placing a print reader and software on every computer station.

Another somewhat promising but less unique personal security tactic is the smart card, a personal identifier card, much like a credit card. This type of security may overcome some human frailties, such as impatience, carelessness, and simple laziness. However, it does not address the problems of lost, stolen, or loaned smart cards. There is no guarantee that the user belongs to the card. This technique has the most promise in a closed LAN situation. It will at least save log-on time and offer some level of security. This system can be enhanced by the user entering a PIN code, much like an ATM card PIN. However, the problem remains of having a card reader at every location from which the user will address the network. This system is called the symmetric key infrastructure (SKI).

14.8 WRITING A BID PROPOSAL REQUEST

It is the responsibility of the TCM to write or have written a concise, detailed bid proposal request (Chapter 15). The document must be exact in terms of work to be accomplished, type of materials and equipment to be used (if either or both are important), time requirements, compa-nies' responsibilities, contractors' responsibilities, and safety responsibilities.

The writer must know exactly what is being requested, so that there is no possibility for confusion on the part of the bidder. If money is the primary concern, specification of a particular manufacturer's products should be avoided. A manufacturer's specification should only be used when no other equipment can possibly perform the tasks.

14.9 SUMMARY

The position of telecommunication manager requires a wide scope of knowledge based on both education and experience. The person in this position should have a strong technical background as well as managerial skills. This is not a position for on-the-job training.

QUESTIONS

1. Who should plan the wiring system?
2. Who should be responsible for tracking the wiring?
3. Who should be responsible for evaluating the cost factors relating to retrofitting?

Writing the Specifications for a Bid Proposal

15.1 INTRODUCTION

Preparing the specifications for an **information for bid** (IFB) is one task that few telecommunication personnel would seek. This task is usually placed on about the same level as that of documentation and layout of the wiring system. Like documentation, this task is very important to the well-being of the communication system. An IFB also has ramifications that are reflected in the morale of the user community and the company's profit and loss statement.

A poorly written IFB can result in incorrect installation of cabling, installation of the wrong type of cable, and/or installation of the cabling at the wrong place. This can cause system down time, poor productivity, or cessation of a user's station or the entire office.

15.2 DETAILS TO INCLUDE IN A REQUEST FOR PROPOSAL

The **request for bid proposal** (RFP) should be written by someone who is attentive to detail. The writer must collect information from all the parties involved in the project so that it is fully understood. This should take place well in advance of the proposal release so that there is time to understand the needs of all parties, write the proposal, have the interested parties review the proposal, and rewrite the document to include any concerns of the affected parties.

The RFP writer should take the following steps in preparing the request.

15.2.1 Research Time

There must be sufficient lead time for the writer to research the project before writing begins. This research should include the following:

1. Contact and/or meet with all interested parties to determine exactly what is needed/wanted and have these needs/wants approved by upper management.
2. Determine the type of cabling that is presently installed in the affected areas and the type of cabling that is required for any new equipment that is included on the approved need list.
3. Determine from the planning department or product vendors the types of cabling and terminations that are necessary for any new equipment.
4. Determine the exact description of any specific products so that these can be specified to the contractor. The writer must be specific or the vendor may install something that meets the specifications of the proposal but not the needs of the users.

15.2.2 Special Facility Consideration

The RFP writers must consider the following:

1. **Equipment rooms:** Are equipment rooms available and are they of sufficient size? It is false economy to have an equipment cabinet stuck in a corner of an office or hall. The needs for connections will grow with time.
2. **Ventilation:** Is there proper ventilation in the equipment rooms and the offices/laboratories to accommodate the new installation?
3. **Lighting and power:** This is an area that is often overlooked. Users seem to believe that systems and equipment can be continually added to the AC voltage outlets or that a new line can be run in from the power panel. Accumulate the power requirements for all the new equipment that is to be installed and present the data to the plant electrical department for a review of the power availability. The cost of any additions should be considered *before* any contract proposal is written. This additional cost may change the needs of the user group. Sometimes the cost of additional power far exceeds the cost of the improvements to the data communication system.

15.3 DEVELOPMENT TIME

The writer must allow sufficient time for the following:

1. Conduct meetings before the actual writing to collect information from the user community.
2. Review the information and perform the actual writing.
3. Have the user community review the completed document to ensure that what they said they wanted is what they really want (probably by now their needs have changed).

15.3.1 Contractor Response Time

Time must be allotted for the following:

1. To get the contract proposal into the hands of the appropriate contractors
2. To allow contractors to review the proposal and respond with bids
3. To review the bids and compare the contractors' responses to the bid proposal

15.3.2 Installation, Testing, and Evaluation Time

A contract would not be written unless the user community had pressing needs for new products. However, telecommunication personnel must allow time for the following:

1. To allow the contractor to perform the actual work
2. To test, evaluate, and accept the installation

15.4 SPECIAL CONTRACTOR CONSIDERATIONS

Before submitting a bid proposal, the writer should review the document from the point of view of the contractors. He or she should ask the following questions:

1. Is the proposal specific on every item that is requested or is any part vague?
2. Does the proposal specify unit pricing where appropriate? Whenever possible, specify separate pricing for material and labor. This will allow you to make a better comparison of the separate bids. Bids that are too low in either area should be suspect.
3. Does the proposal specify a deadline for bid response? A bid response time should be specified with date, year, time of day, and place (for example, October 1, 1997, at the Office of the President, Somerset Press, 5200 Dream Ranch Circle, Somerset, CA 95667). Specify that it is the responsibility of the contractor that mailed bids meet the time deadline.

15.5 BID PROPOSAL FORMS

You may wish to develop specific forms for both requests for proposal and bid response. Figure 15-1 is a sample of such a form. The forms can be rather simple or contain great detail. Figures 15-2 through 15-9 are examples of very detailed forms for part of an IBM work request.

A sample of a step-by-step detailed work proposal is presented in the following section.

15.6 DETAILED WORK PROPOSAL

The detailed work proposal (DWP) is a longer format that is preferable to the RFP. The following paragraphs will detail the main parts of that type of document.

The purpose of such documents is to provide vendors with the specific information and requirements (discussed earlier) for job estimates, including materials and labor.

REQUISITION #				CURRENT DATE

REQUESTOR NAME	DEPT.	WK PH	WORK *o CODE	

NEED DATE	ORDER NUMBER	JOB/PROJECT NUMBER

TYPE OF ACTIVITY REQUESTED *1

DESCRIPTION OF PURCHASE

ACTIVITY	PARTS REQ'D	PART #	QUANTITY	UNIT PRICING	TOTAL
ITEM 1					
2					
3					
4					
5					
6					
7					
TOTAL (ESTIMATE)					

*o WORK CODE

701 = ENGINEERING
702 = TEST
703 = R & D
704 = SPECIAL PROJECT
705 = SUPPORT

ACTIVITY *1

501 = PRODUCT INSTALLATION
502 = SERVICE REQUEST
503 = PRODUCT & SERVICE REQUIREMENTS
504 = OUTSOURCE LABOR
505 = VENDOR

APPROVAL SECTION

MANAGEMENT LEVEL 1: _____ NAME _____ PH _____ JOB CODE _____ AUTHORIZATION LEVEL

MANAGEMENT LEVEL 2: _____
MANAGEMENT LEVEL 3: _____

ACCOUNTING SECTION

IF CAPITAL PURCHASE: ( ) CAPITAL   COST _____ (REQUIRED)
                                     BUDGET _____ (BY DEPT./FUNCTION)

                     ( ) NEW   ( ) E/C
                     ( ) EXPENSE   COST _____
                                     BUDGET _____

                     ( ) OTHER _____ (REQUIRED IF CHECKED)

ACCOUNT APPROVAL

/BUDGET
APPROVER/PLANNER _____ (NAME) EXT _____
ACCT. NUMBER: _____
CAPITAL PLAN/PROJECT: _____

( ) FIXED ASSETS:
( ) ACTUAL TOTAL COST OF REQUEST $ [   ]

VERIFIED NO _____
          YES _____

HIGHEST LEVEL
OF MANAGEMENT APPROVAL: _____

NAME _____ / _____ DATE

SAFETY HAZARDS

PERSONNEL   ( )
CHEMICAL   ( )
ELECTRONIC ( )
ELECTRICAL ( )
FACILITY   ( )

Figure 15-1 RFP form.

Attaching Products Worksheet			
Accessories	Part Number	Total Number	Comments
Multiuse Communication Loop			
MCL-1 Type 1LS Loop Station Connector (LSC)	4760511		
MCL-2 Loop Wiring Concentrator (LWC)	6091077		
MCL-3 Component Housing	6091078		
MCL-4 Cable Bracket	6091042		
MCL-5 Patch Cable (8 feet)	8642551		
Series/1			
S/1-1 MFA/422 Attachment Cable	8310553		
S/1-2 Y Assembly	8642549		
S/1-3 Twinaxial Y Assembly	8642550		
S/1-4 Twinaxial Straight Adapter	7362230		
S/1-5 Patch Cable (8 feet)	8642551		
S/1-6 Patch Cable (30 feet)	8642552		
S/1-7 Series/1 Feature #5790	- - - - - - -		
S/1-8 Twinaxial Impedance Matching Device	6091070		
S/1-9 Twinaxial Terminator	6091068		
S/1-10 Twinaxial Direct Connect Cable	6091075		
5080 Graphics			
5080-1 Red Coaxial Balun Assembly	8642546		
5080-2 Single Cableless Balun Assembly	6339082		
5080-3 Double Cableless Balun Assembly	6339083		
5080-4 Y Assembly	8642549		
General Purpose Attachment			
Gen-1 General Purpose Attachment Cable	8310554		
Gen-2 Patch Cable (8 feet)	8642551		
Gen-3 Patch Cable (30 feet)	8642552		

Figure 15-2   DWP form.

The following format will assure the concise, complete work proposals that are essential for accurate bidding purposes.

For large, complex, or special work assignments, corporations may require an extra layer of administration (namely, a select committee that will select the best vendor or contractor to be awarded the work).

Be mindful of the bidding process and what it implies. An extremely low bid may be submitted by a vendor to attempt to obtain future business. Very low bids may not contain a reasonable amount of profit. This may indicate that in an attempt to get the present work assignment, the contractor may be looking for ways of cutting cost outside of specifications.

Complete Order Summary Worksheet (Part 1 of 4)

Cables: For installation and maintenance, order 15% additional cable.

Type	Part Number	Meters (feet)
1	4716748	
1 Plenum	4716749	
1 Outdoor	4716734	
2	4716739	
2 Plenum	4716738	
5	4716744	
6	4716743	
8	4716750	

Equipment Racks: Racks are not available from IBM. Order from your electrical supplier or contractor. Racks may not be a stock item, so allow enough lead time.

Type	Quantity
Open Rack	
Enclosed Rack	

Accessories:

Description	Part Number	Quantity
Cable Tester Kit (includes tester, case, data wrap plug, and batteries)	4760500	
Cable Tester (includes batteries)	4760501	
Twinaxial Test Accessories (includes twinaxial test adapter, twinaxial test terminator, and two twinaxial straight adapters	6339087	
Telephone Tester Attachment Kit	4760509	
Data Wrap Plug	4760507	

Note: For large installations where extensive tester usage is anticipated order:
- One 8-foot patch cable
- Additional data wrap plugs

This will extend the life of the data test cable connector and also facilitate testing multiple offices from the wiring closet.

Figure 15-3 Complete order summary worksheet (page 1).

Complete Order Summary Worksheet (Part 2 of 4)		
Accessories: For installation and maintenance, order 10% additional accessories. Order at least two additional surge suppressors of each type used.		
Description	Part Number	Quantity
Data Connector *	8310574	
3-Pair Telephone Jack *	8310575	
3- or 4-Pair Telephone Jack *	8310551	
Type 1 Faceplate *	8310572	
Type 1 Faceplate for Japan *	6339094	
Type 2 Faceplate for 3-Pair Telephone Jack *	8310573	
Type 2 Faceplate for 3- or 4-Pair Telephone Jack *	6091025	
Type 2 Faceplate for 3- or 4-Pair Telephone Jack for Japan *	6339095	
Type 1W 87mm *	6091048	
Type 1W 80mm *	6091049	
Type 1S Surface Mt	4760486	
Type 2S Surface Mt for 3-Pair Telephone Jack	4760485	
Type 2S Surface Mt for 3- or 4-Pair Telephone Jack	6091029	
Distribution Panel	8642520	
Rack Ground Kit	4716804	
Surge Suppressor	4760469	
Cable Loc Chart	4716816	
Cable ID Label (8 sheets)	4716817	
Undercarpet Cable Connector Kit* (Note 3)	6339123	
Floor Monument (Note 3)	6339128	
Floor Monument Faceplate Kit (Note 3)	6339131	
Undercarpet Cable Wall Box (Note 3)	6339130	

Note:
* Can only be ordered in multiples of 25
3. Not available from IBM

Figure 15-4 Complete order summary worksheet (page 2).

```
┌─────────────────────────────────────────────────────────────────────┐
│ Complete Order Summary Worksheet (Part 3 of 4)                        │
│  ┌──────────────────────────────────────────────────────────┐        │
│  │ Accessories: For installation and maintenance,            │        │
│  │          order 10% additional accessories.                │        │
│  └──────────────────────────────────────────────────────────┘        │
│         Accessories used in more than one application                 │
```

Description	Part Number	Quantity
Loop Wiring Concentrator (LCW)	6091077	·
Cable Bracket	6091042	
Red Coaxial Balun Assembly	8642546	
Single Cableless Balun Assembly (note 3)	6339082	
Double Cableless Balun Assembly (note 3)	6339083	
Y Assembly	8642549	
Twinaxial Y Assembly	8642550	
Twinaxial Impedance Matching Device	6091070	
Twinaxial Direct Connect Cable	6091075	
Twinaxial Terminator	6091068	
Patch Cable (8 feet)	8642551	
Patch Cable (30 feet)	8642552	
General Purpose Attachment Cable (note 1)	8310554	
Coaxial Accessories		
Coaxial Patch Panel	4176801	
Yellow Coaxial Balun Assembly	8642544	
Single DPC Attachment Cable (8 feet)	6339073	
Single DPC Attachment Cable (20 feet)	6339074	
Double DPC Attachment Cable	6339075	
3299 Mounting Shelf	6217036	
Spare BNC Bulkhead Connector	(note 2)	
Twinaxial Accessories		
Twinaxial Test Accessories Kit	6339087	
Finance Communication Loop Accessories		
Plug and Jack Assembly	8310552	
Y Assembly	8642549	
Store System Loop Accessories		
WE Type-404B Receptacle	(note 3)	

```
Notes:
1. Can be ordered for use as data wire test cable
2. Not available from IBM. Order Amphenol 31-220 or equivalent.
3. Not available from IBM.

Continued
```

Figure 15-5 Complete order summary worksheet (page 3).

Complete Order Summary Worksheet (Part 4 of 4)

Accessories: For installation and maintenance,
order 10% additional accessories.

Accessories used in more than one application

Description	Part Number	Quantity
Multiuse Communication Loop Accessories		
Type 1LS Loop Station Connector (LSC)	4760511	
Loop Wiring Concentrator (LWC)	6091077	
Component Housing	6091078	
Series/1 Accessories		
MFA/422 Attachment Cable	8310553	
Twinaxial Straight Adapter	7362230	
Series/1 Feature #5790	-------	

Figure 15-6   Complete order summary worksheet (page 4).

Rack Inventory Chart

Wiring closet number _____

Rack number _____

Date _____

Planner's initials _____

Instructions

Fill out a Rack Inventory Chart for each equipment rack.

1. Enter the wiring closet location number, the equipment rack identification number, and the planner's initials.

2. Using the template for the Rack Inventory Chart that came with this manual, draw an outline of each component that will be installed in the rack.

3. The slots at the bottom of the distribution panel template are used only for the lowermost distribution panel in a rack. The slots indicate that there are 38.1 mm (1 - 1/2 in.) between that panel and the next unit in the rack.

4. Write the unit identification number on each component on the chart.

Example:

	21
	22
0010	
0011	
0012	

Figure 15-7 Rack inventory chart form.

System Configuration Worksheet

System   **Service Contact**   **Telephone**

Attachment Description	Accessories in Work Area	Cable Runs from (Wall)	Cable & Cable Length	Cable Runs to (Panel)	Accessories on Distribution Rack

Suggested Accessory Abbreviations

GPA — General Purpose Attachment Cable	RCB — Red Coaxial Balun	LSC — Loop Station Connector
MFA — Multifunction Attachment Cable	SCB — Single Cableless Balun	LWC — Loop Wiring Concentrator
Y — Assembly	DCB — Double Cableless Balun	PJ — Plug and Jack Assembly
TY — Twinaxial Y Assembly	YCB — Yellow Coaxial Balun	AD — Adapter
IMD — Impedance Matching Device	SDPC — Single DPC Attachment Cable	P — Patch Cable
TDC — Twinaxial Direct Connect Cable	DDPC — Double DPC Attachment	CPP — Coaxial Patch Panel
		SS — Surge Suppressor

Figure 15-8   System configuration form.

Wiring Closet/Controller Room Worksheet

Building _____
Floor _____
Worksheet _____

Cable Routes Within a Single Building

	Wiring Closet Location/ Floor	Wiring Closet or Controller Room Location/ Floor	Number of Cables	Cable Length	Cable Requirements					
					Type 1	Type 1 P	Type 5	Faceplate Devices		
								1	1S	1W
1										
2										
3										
4										
5										
6										
7										
8										
9										
10										
11										
12										
13										
14										
15										
Totals										

Cable Routes Between Buildings

	Wiring Closet Location/ Floor	Surge Suppressor Location/ Floor	Wiring Closet or Controller Room Location/ Floor/ Building	Length of Indoor Cable in this Building	Cable Requirements								
					Type 1		Type 1 P		Length of Outdoor Cable	Type 1 Outdoor		Surge Suppressors
					No.	Total Feet	No.	Total Feet		No.	Total Feet	
1												
2												
3												
4												
Totals												

Data Connectors _____

Distribution Panels _____
Distribution Racks _____

Rack Grounding Kit _____
Cable Label Packages _____

Figure 15-9 Wiring closet/controller room worksheet.

The following is a general outline for a DWP:

Section 1   General specifications
Section 2   Definitions and contracts
Section 3   General installation guidelines
Section 4   Work assignments
Section 5   Equipment and materials list
Section 6   Installation specifications
Section 7   Architectural and layout drawings
Section 8   Testing and debugging procedures
Section 9   Database updates and revisions
Section 10 Quality control checkpoint
Section 11 Corporate acceptance tests
Section 12 Vendor warranty

15.6.1   Section 1 General Specifications

1. *National compliance* All work will be done in compliance with the national, state, and local codes (including, but not limited to, building, electrical, fire, and health).

2. *Corporate specifications* All materials (cabling, electrical, electronic, etc.) must meet corporate specifications as outlined in this document.

3. *Vendor prior work experience* The vendor awarded this contract shall have prior experience with this type of installation or work activity. Proof of the prior work shall be made available upon request.

4. *Corporate compliance* All work shall be in compliance with the following corporate standards and publications: [In this section, if the work to be performed is to meet an existing corporate criterion, all references should be noted and availability and access resources should be noted, so that the contractors can review the various criteria. If no additional criteria are required, this section should be omitted.]

5. *Cleanup and damage* The contractors shall be responsible for cleanup in all areas in which work is performed, including reinstallation of all equipment disturbed during the work activity. The contractor is also required to replace ceilings, floor tiles, walls, floors, etc. damaged during work activity.

6. *Service interruption and reconnection* Vendors shall be responsible for the reconnection of any communication, telephone circuitry, and/or power disrupted during the work activity.

7. *Time/target schedule* A schedule showing a time line listing all work by period divisions will be provided. Specific targets will be monitored by time/day acceptance by the Project Coordinator.

8. *Access* Details of how the vendor will gain access to the work area should be presented here. If the area is a secured area, plant security will also have to be notified as to who will be needing access and for how long.

9. *Work impact* While the work activity is in progress there may be interruptions to the communications network, computing system, building power, lighting systems, etc. All "impacted areas" need to be identified and all affected departments must be notified well in advance so that they can adjust their work schedule.

10. *Labeling* All cabling runs, conduits, terminal blocks, and so on shall be labeled in compliance with all corporate and required standards.

15.6.2 Section 2 Definitions and Contracts

This section is for application-specific definitions of common word usage. The following are definition examples:

Company: This is the common name as referred to throughout this document for the requesting company.

Cable 1mm: Plenum cable 93 Ω fully connectorized with male-to-male connections.

Cable 2: Ethernet thin cable.

Cable 3: Fiber-optic cable pair.

Outlet 2: IBM cable type 1 connectorized wall plate plus one RJ-11 type 3 phone jack.

Premise wiring system: Existing wiring and cable media systems.

The following section is for work-related contracts. Examples follow:

Projector Coordinator: Lead person designated to coordinate all work activities and groups.

Name: _____ Day Phone: _____

Support Service Contact: In the checklist of affected groups there should be a hardware and software contact as well as a system contact person. All phone numbers should be given and backup personnel should be listed.

Hardware Support: Name: _____ Phone: _____

System Support: Name: _____ Phone: _____

15.6.3 Section 3 General Installation Guidelines

This section is for *installation specifications* and requirements. For example, if you are using this document to obtain an installation quote on voice, data, teleco, ISDN, or fiber cabling, this section will assist the requester to define such items as the following:

1. Correct cable type and impedance rating.

2. Proper terminations for each cable group.

3. Minimum and maximum cable length in work areas.

4. Appropriate outlet fixtures for office, lab, or other work areas.

5. Proper labeling consistent with corporate and national standards.

6. Cable pathing. Reference premise cable runs and layout. Ensure that all runs are within cable trays where specified.

7. Reference exhibits for all cable trays and cable runs.

If any component of the work must adhere to standards (e.g., corporate), this would be the area in which to refer to such standards by publication number or index resource.

15.6.4 Section 4 Work Assignments

In this section the actual work description will be given. The following is an example of typical work assignments.

- Install IBM cabling system rack for use with IBM 3270 products on computer floor 3 location j4-18.
- Port wiring harness (coax) from control unit (3×) shall be connected directly to back of coax patch panel.
- Hardware required at computer floor:
    - 1 equipment rack
    - 1 distribution panel
    - 4 24 BNC female/female coaxial patch panels
    - 1 3299 multiplexer
    - 1 equipment shelf
    - 1 interconnection type 1, 2, or 9 cable run between back of distribution panel to office area
    - 10 cable guides
    - 20 interconnection cables
    - 10 jumpers
- Hardware required at office:
    - 1 type 1 face plate
    - 1 single cable coaxial balun assembly (which can be used instead of red balun assembly; (93 to 150 Ω converter)
    - 1 coax cable
    - Please note that all required hardware and equipment lists are provided in Section 15.6.5.
- Installation details are covered in Section 15.6.6.

15.6.5 Section 5 Equipment and Materials List

Description	Part Number	Quantity	Source	Comments
Equip Rack	8899xx-2	1	Vendor	
Equip Shelf	0008989	2	Vendor	
Distrib Panel	5555555	1	Vendor	
3299-1Multiplexer	0xxxx78	1	Vendor	
Coax Patch Panel	9090.8	4	Vendor	
Cable (IBM type 1)	CCCxxx	2000 ft	Vendor	
Cable Connectors	909090	50	Vendor	
Baluns	1m9898	6	Vendor	
Coax (93 Ω)	NNNNN	400 ft	Vendor	
Cable Guides	poppp8	20	Generic	
Coax Terminations	1787	50	Vendor	
Face Plates	87877n	1	Vendor	

15.6.6 Section 6 Installation Specifications

Assemble IBM cabling system and install rack in location A as shown on *Layout 1*, including orientation.

Once all mechanical hardware is assembled, begin recabling process. *Note:* Below tiles marked "FF" is the controller port coax harness. Locate and connect to coax patch panels at bottom of equipment rack through cable-access cut floor tile. All other connections and links will be performed by company personnel.

15.6.7 Section 7 Architectural and Layout Drawings

The following layouts are for work assignments 1 and 2.

15.6.8 Section 8 Testing and Debugging Procedures

The contractor is to perform all first- and second-level problem determination on all vendor-installed equipment and services. Vendor shall provide all necessary test equipment and personnel to ensure that adequate testing and debugging is an internal part of any testing procedure.

Contractor is to ring out all single- and multicable bundles once connectorized. All high-resistance and open runs will be corrected prior to release back to Project Coordinator.

15.6.9 Section 9 Database Updates and Revisions

After all moves, adds, and changes are completed, the Project Coordinator shall be responsible for getting the database updated to reflect current conditions. This task can be delegated to database management.

15.6.10 Section 10 Quality Control Checkpoint

A person from Corporate, preferably from Quality Control, shall be designated to ensure that all major tasks are completed in accordance with existing guidelines. A checklist should be provided that begins with inventory, work activities, and so forth, including the final step of database updating.

15.6.11 Section 11 Corporate Acceptance Testing

Vendor shall notify Project Coordinator of work assignment competition. The appropriate corporate function will then perform mandatory acceptance testing. Any problems noted during these acceptance tests will be reported to Vendor for correction. Time is of the essence. All corrections are to be made within an acceptance time limit as prescribed by corporate directive.

Upon satisfactory acceptance testing of work assignment a corporate acceptance letter shall be issued.

15.6.12 Section 12 Vendor Warranty

Vendor shall warranty all materials and labor for a minimum of 1 year from date of corporate acceptance.

15.7 SUMMARY

The proposal for bid submissions should be written only after a detailed study of the project, and then by a person who is committed to detail. Finally, the proposal should be reviewed by the department "nit picker" before being submitted to contractors.

QUESTIONS

1. Why is it important that the specifications for the call bids be detailed?
2. When should a low bid be suspect?
3. Should a corporation expect every contractor to have experience in the area of a bid?
4. Who is usually responsible for testing and troubleshooting a cabling installation?
5. Who performs the final acceptance test?
6. When does the vendor's guarantee begin?

CHAPTER 16

New Technology

16.1 INTRODUCTION

Communication technology seems to be increasing exponentially. Each new idea and innovation tends to bring on ten more. While many new ideas and concepts do not meet marketplace acceptance, many flow into the mainstream. DSL (Chapter 7) is one technology that seems to meet the marketing need of faster and more economical information transfer. It is inevitable that new technologies will emerge. Only time will determine their acceptability.

16.2 POWERLINE TECHNOLOGY

The PowerLine idea is so simple—transmission of data over the electrical power grid—that we will all say, "Why didn't I think of that?" In America and in most of the developed world, there are power outlets in every room in every home, school, and commercial building and outside the buildings. If HomePlug can overcome the problem of power line electromagnetic noise spikes from dishwashers, hair dryers, lightning strikes, power surges, and varying voltage levels, the Internet can be connected to any room or outside power outlet to form a network. Examples of the items developing noise on home power lines are illustrated in Figure 16-1. The questions involve how this can be accomplished and who will profit.

The how part was approached on June 5, 2000, by an organization called the *HomePlug Powerline Alliance* based in San Ramon, California. The alliance, comprised of 36 companies including Intel, Hewlett-Packard, 3Com, Cisco Systems, AMD, and Tandy/Radio Shack, selected a technology standard that will serve as a common way for connecting electronic devices to the Internet through electrical outlets. Among the baseline technologies were the ability to handle 10 Mbps with high security and to handle devices connected into a large number of power outlets. The alliance chose *orthogonal multiplexing frequency division* technology by

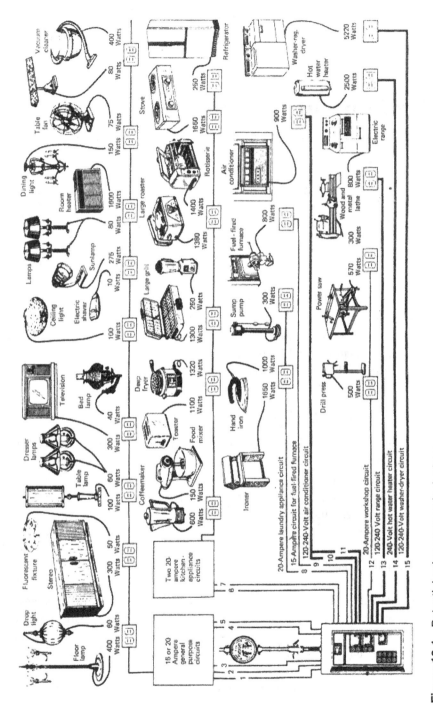

Figure 16-1   Potential noise sources in the home.

Intellon Corp. of Ocala, Florida—specifically, PowerPacket—as the transmission technology. PowerPacket has a raw 14-Mbps throughput and low noise interference. This technology is supposed to pull data out of the extremely noisy environment of electrical power.

Intellon says that "the technology will work with DSL, DSL Lite, and asymmetric DSL and won't interfere with power line technologies such as X-10 used to control household devices, because the technologies operate on different carrier frequencies."

The HomePlug alliance conducted field trials in 500 homes during the sumer of 2000. Their expectations are to have products on the market in 2001. Universality and low cost are expected to be to main selling points in marketing PowerLine products.

The first of the PowerLine products available will be modemlike devices that connect Internet phone lines and/or DSL lines to in-house wiring to form a building network.

HomePlug is attempting to override 802.11b 11-Mbps wireless technology, which has the advantage of allowing workers to roam anywhere within the premises. HomePlug is also bucking HomePNA, through which every plug in a building becomes a 1.0-Mbps port. The PowerLine technology is expected to jump to 10 Mbps when Version 2.0 emerges.

16.3 SUMMARY

PowerLine technology has great possibilities for inexpensive home networks. However, with the number of organizations working on a variety of solutions and phone-based DSL and wireless ahead in the game, it remains to be seen if PowerLine technology will be a big factor in the market.

Figure 16-2 depicts a coal-fired generator and power distribution system. The example demonstrates the problems associated with tagging the Internet on power lines to residences or offices.

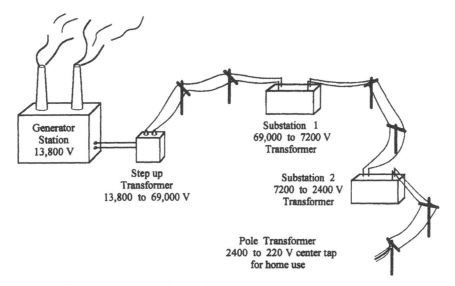

Figure 16-2 A power distribution system.

GLOSSARY

A   Ampere. A measurement of current. (See *Ampere*.)

AAL5   ATM Adaption Layer 5. AAL5 has been adapted by the ATM Forum for a class of service called high-speed data transfer.

Abrasion Resistance   Ability of a wire or cable to resist surface wear.

Access Line   The connection between the subscriber's facility and the public network.

Access Rate   The bit per second (bps) rate at which a user can transmit over the lines that access the network.

ACR   Attenuation cross talk ratio. The difference between attenuation and cross talk, measured in decibels, at a given frequency. Important characteristic in networking transmission to ensure that signals sent down a twisted pair are stronger at the receiving end of the cable than is any interference imposed on that same cable pair.

Adapter   A mechanical media termination device designed for cable connections. Sometimes referred to as coupling, bulkhead, or interconnection devices.

ADSL   Asymmetric digital subscriber line. An ANSI standard that provides a voice connection plus a data connection on the same customer local loops that currently support only one analog voice connection. The data connection supports from 64 Kbps to 640 Kbps upstream (customer network) and 1.5 Mbps to 8 Mbps downstream. (See *DSL*.)

AES/EBU   Informal name for a digital audio standard established jointly by the AES (Audio Engineering Society) and EBU (European Broadcast Union).

AF   Audio frequency.

Air-Gap Dielectric   A coaxial design in which a monofilament of plastic holds the center conductor in place, allowing the remainder of the dielectric to be air. Typically, velocities of up to 80% can be achieved in this design.

AIX   Advanced Interactive Executive. IBM's implementation of UNIX.

Alloy   A combination of two or more polymers/metals. Usually combined to make use of a different property of each polymer metal.

ALOHA   An experimental LAN of the packet-switching type utilized by the University of Hawaii.

Alternating Current (AC)   Electrical current that alternates or reverses polarity continuously. The number of alternations per second is called Hertz (Hz).

Altos™ Cable   Siecor™ loose tube cable in which buffer tubes contain two or more fibers and that uses Dry™ water-blocking technology for craft-friendly installation.

AM   Amplitude modulation.

Ambient   Conditions existing at a test or operating location prior to adding heat or cooling sources (i.e., ambient temperature).

American Wire Gauge (AWG)   A standard for expressing wire diameter. As the AWG number becomes larger, the wire diameter becomes smaller.

AMI   Alternate mark inversion. A line coding technique used to accommodate the ones density requirements of E1 and T1 lines.

AM Modulation   One of the methods of inserting information into a carrier. In this case, the amplitude of the carrier is varied at the rate of the carrier signal.

Ampere   A standard unit of current. Defined as the amount of current that flows past a given point when 1 V is applied across a resistance of 1 Ω. An ampere is produced by 1 C of charge passing a point in one second, 6.25×10^{18} electrons past a point in one second.

Amplifier   An electronic device used to boost (amplify) a signal.

Amplitude   The maximum value of a varying waveform.

Amplitude Distortion   The unwanted change of the amplitude of a signal. This distortion is sometimes called noise.

AN   Access node. (See *Node*.)

Analog   A format that uses continuous physical variables, such as voltage amplitude or frequency variations, to transmit information.

Analog Signal   An electrical signal that varies continuously and does not have discrete values. Analog signals are copies of other waveforms in nature.

Anneal   To soften and relieve strains in any material, such as glass or metal, by heating and slowly cooling the material. Annealing generally lowers the tensile strength of the material and improves its flexibility.

ANSI   American National Standards Institute. A U.S. standards organization, ANSI cooperates with governmental, industry, and vendor groups and associations to develop voluntary standards in many areas, including communications. ANSI also participates in the International Standards Organization (ISO) to develop voluntary international standards.

Application Layer   The seventh and highest level in the open systems interconnections (OSI) model that contains the user application programs.

Aramid Yarn   Strength member element used in Siecor cable to provide support and additional protection of the fiber bundles. Kevlar is a particular brand of aramid yarn.

Architecture   The structuring of a LAN or the operation of a computer.

Armoring   Additional protection between jacketing layers to provide protection against severe outdoor environments. Armoring is usually made of plastic-coated steel and may be corrugated for flexibility.

ASCII   The American Standard Code for Information Interchange, an 8-bit code used for information exchange. Seven bits represent the characters and the eighth bit is used for parity.

ASIC   Application-specific integrated circuit. A hardware technology in which integrated circuits are manufactured for customized applications to reduce space for speed and security reasons.

ASTM   The American Society for Testing and Materials, a standards organization that suggests test methods, definitions, and practices.

Asynchronous Transfer   An approach for the efficient transfer of information, in which time slots are used on a demand basis (e.g., by STDM, ATM) rather than on a periodical (synchronous) basis (e.g., TDM, STM).

Asynchronous Transmission   A transmission method in which the time intervals between characters may be of unequal length. Asynchronous transmission is sometimes called start-stop transmission.

ATM   Asynchronous transfer mode. A broadband switching and multiplexing, connection-oriented, high performance integrated technology for supporting B-ISDN services (multi-media) under specified QoS guarantees. Since capacity is allocated on demand with no clocking control between users, it is called asynchronous. Information is transferred at a very high rate in fixed-length packets called cells. Traffic streams can be distinguished according to different QoS classes. The ATM Reference Model describes the various functions, services, protocols, and standards encompassed by ATM technology.

Attenuation   The decrease in magnitude of power of a signal in transmission media (copper, coax, fiber, or hardware) between points. A term used for expressing the total loss in an optical fiber consisting of the ratio of light output to light input. Attenuation is usually measured in decibels per kilometer (dB/km) at a specific level. Attenuation is also measured as the logarithm of a ratio and expressed in decibels (dB).

Attenuation Coefficient   The rate of optical power loss with respect to distance along the fiber, usually measured in decibels per kilometer (dB/km) at a specific wavelength. The lower the number, the less the fiber attenuation. Typically, multimode wavelengths are 850 and 2300 nm; single-mode wavelengths are 1310 and 1550 nm. When specifying attenuation it is important to specify whether the value is average or normal.

AT&T   American Telephone and Telegraph, Inc.

ATU   ADSL transceiver unit. The ADSL Forum uses terminology for DSL equipment based on the ADSL model for which the forum was originally created. The DSL endpoint is known as the ATU-R and the CO unit is known as the ATU-C. These terms have since been used for other types of DSL services, including RADSL, MSDSL, and SDSL. ATU now represents xDSL services.

ATU—Central Office   The ATU at the carrier's central office in support of DSL-based services.

ATU—Remote   Equipment placed at the customer's premises in support of DSL-based services.

Audio   A term used to describe sounds within the range of human hearing. Also used to describe devices designed to operate within this range of 20 to 40 kHz.

AWG   American wire gauge. A measurement of wire diameter and area in circular mils. The smaller the number, the larger the wire size, and therefore the lower the resistance and greater the current-carrying capacity.

AWM   Appliance wiring materials.

Backbone   Equipment that provides connectivity for users of a distributed network, including the network infrastructure.

Backbone Cabling   That portion of premise cabling providing connections among telecommunications closets, equipment rooms, and entrance facilities. Backbone cabling consists of transmission media, main and intermediate cross connects, and terminations for the horizontal cross connect, equipment rooms, and entrance facilities. Backbone cabling can also be classified as campus backbone (cabling between buildings) or building backbone (cabling between floors or closets within a building).

Backbone LAN   A transmission facility designed to connect two or more LANs.

Backbone Network   The core communications system that provides connectivity between communities of users. It may span a campus or a business park (campus backbone) or a larger geographic area such as a city or a county, or span international boundaries (WAN backbone).

Balanced Line   A cable having two conductors that carry voltages opposite in polarity and equal in magnitude with respect to ground.

Baluns   Impedance-matching devices comprised of a transformer that connects a balanced line (twisted pair) to an unbalanced line such as a coaxial cable.

Bandwidth   Measure of the information-carrying capacity of any communications media. For analog circuits, bandwidth is measured in Hertz (Hz or cycles per second). For digital circuits bandwidth is measured in bits per second (bps).

Bandwidth-Distance Product   The information-carrying capability of a transmission medium, normally referred to in units of MHz•km. This is called the bandwidth-distance product or, more commonly, bandwidth. The amount of information that can be transmitted over any medium changes according to distance. The relation is not linear. A 500 MHz•km fiber does not translate to 250 MHz•km for a 2-km length or 1000 MHz for a 0.5-km length. When comparing media it is important that the same units and distance are used.

Baseband   A frequency band occupied by a single signal in its original unmodified form.

Baseband Transmission   A transmission method that uses low frequency starting at zero Hertz and carries only one transmission at a time. This method utilizes no carrier.

Battery Backup   A battery that supplies power to a system in the event of the loss of main power.

Baud   The rate at which signals are transmitted. A bit rate of 2000 is a Baud rate of 2000.

Bel   A unit that represents the logarithm of the ratio of two levels. The unit is of such large value that the decibel is most often used to describe the ratio of two voltage, current, or power levels. The dB is sometimes used to describe hydraulic levels. dB = $10 \log P_1/P_2$, $20 \log E_1/E_2$, and $20 \log I_1/I_2$.

Bend Loss   The energy loss caused by having an optical fiber curved around a restrictive radius of curvature.

Bend Radius   The amount that a fiber can bend before the risk of breakage or increase in attenuation. Also can refer to the maximum bend of a coaxial cable.

BER   Bit error rate. Measure of transmission quality indicating the number of bits incorrectly transmitted in a given bit stream compared to the total number of bits transmitted in a given time.

Biconic   A type of connector for a fiber optic cable.

Binary   The two states of a digital system comprised of ones and zeros or signals and no signals.

Binder   A tape or strap used to hold a cable in place.

Bit   A binary digit, the fundamental unit of information expressed in digital form. The choice between two states such as yes or no, high or low voltage, or off or on.

Bit Stream   A series of digital pulses or a digital signal.

Bits per Second (bps)   The number of bits of digital signal per second. To obtain the bit rate of a modem, multiply the number of signal changes per second (baud rate) by the number of bits of information carried by each change.

BNC   A type of connector with a bayonet-type locking mechanism used to terminate coaxial cable.

Bonding   The connection of all building equipment electrical grounds to eliminate any difference in ground potentials.

Booster   An amplifier or other device inserted into a line or cable to increase voltage or power. Transformers are used to boost voltage or current, and amplifiers are used to increase power in signal lines or at antennas.

BPS (Bits per Second)   The number of bits passing a point per second. A measure of the speed of digital transfer rate along a circuit.

Braid   The weaving of metal or textile fibers to form a tubular structure that may be applied over one or more wires or flattened to form a strip.

Breakdown Voltage   The voltage at which the insulation between conductors will fail and allow current to flow between the conductors.

Breakout Box   A device used for testing a circuit or cable.

BRI   (Basic Rate Interface) The ISDN term that refers to the basic ISDN interface of 25 + D. Operating at 144 kbps, BRI provides two B (bearer) channels at 64 kbps and a D (data) channel at 16 kbps.

Bridge   A device that connects two LANs so a node on one can communicate with a node on another.

Broadband   The original term came from analog transmission systems and referred to the capacity of an analog circuit as measured by the range of frequencies it could support within a –3-dB power level. This was called the bandwidth. Broadband referred to circuits that were capable of passing analog signal frequencies greater than 4 kHz, often called narrowband. When digital transmission systems were developed in the 1960s, the term *bandwidth* came to be used to describe the capacity of such circuits, although bit rate or capacity is more accurate. Broadband now refers to digital circuits and transmission lines that are capable of carrying very high bit rates greater than that of ISDN PRI (1.5 Mbps DS-1 or 2 Mbps E1). It is possible that this boundary is becoming too low.

Buffer   A temporary storage device made up of logic circuits. Such a device is used to isolate one system from another in which the data flow, sequence of events, or impedance is different.

Buffering   (1) A protective coating extruded directly on the fiber coating to protect it from the environment; (2) extruding a tube around the coated fiber to allow isolation of the fiber from stress on the cable.

Building Backbone   The portion of a backbone cable within a building (floor to floor or closet to closet).

Bunch Strand   Conductors twisted together with the same lay and direction.

Bundle   Many individual fibers or other cables within a single jacket or buffer tube. Also a group of buffered fibers distinguished in some fashion from another group in the same cable core.

Bus   A common set of cables for a multiuser computer system.

Bus-Bar Wire   Uninsulated tinned copper used as a common lead.

Bus Topology   A physical layout of a LAN in which all nodes are connected to a single cable.

Byte   A collection of 8 bits of information operated as a unit. Most character sets use one byte per character. The capacity of storage devices is often given in bytes or in kilobytes (kbytes) per second, meaning 1024 bytes or 8192 bits per second.

Cabinet   An enclosure for rack-mounted equipment.

Cable A multistrand round or flat bundle of either stranded or solid wires formed into one unit, usually covered by an insulating jacket.

Cable Assembly. An assembly of wires or cables into a single unit with or without connectors. (See *Patch Cord*.)

Cable Bend Radius During installation the term implies that the cable is experiencing a tensile load. Free bend implies a lower allowable bend radius since it is at a condition of no load.

Cable Binder A device used to bundle multiple insulated copper pairs together in the telephone network.

Cable Through The capacity of an information system that allows multiple workstations to attach to a single cable.

Cable Tray A multiconductor cable, designed per the NEC, to be used in a wiring tray.

Cabling The method by which a group of conductors are assembled together or twisted together.

CAD Computer aided drafting.

Call A request for a connection to the system.

Camp-On A PBX or LAN operation that allows a caller (user) to wait on line if the called station is busy.

Campus Backbone The portion of backbone cabling between buildings. (See *Backbone Cabling*.)

Canadian Electrical Code (CEC) Canadian version of the American National Electrical Code (NEC).

Cap Carrierless amplitude and phase modulation. A transmission technology for implementing a DSL connection. Transmit and receive signals are modulated into two wide-frequency bands using passband modulation techniques.

CAP Competitive access provider.

Capacitance The property of an electrical conductor that permits the storage of energy in the form of a displacement of an electrical charge in coulombs. Capacitance is usually measured in picofarads (pF) or microfarads (μF).

Capacitive Reactance (X_C) The opposition to an electric current due to the capacitance in a circuit or a cable. It is measured in ohms. $X_C = \frac{1}{2}\Pi fC$.

Capacitor Two metal surfaces separated by an insulating dielectric. The capacitance of a capacitor is determined by the area of the metal plates and the thickness and type of dielectric material.

CAT5 Category 5. A level of unshielded-pair wiring performing as defined by EIA/TIA-568 (up to 500 MHz).

Category Rating Categories are transmission performance criteria defined by the Telecommunication Industry Association (TIA) for cabling, connectors, and installed cabling systems.

> Category 3—up to 16 MHz
> Category 4—up to 20 MHz
> Category 5—up to 100 MHz
> Category 6—up to 200 MHz
> Category 7—Up to 600 MHz

CATV Cable or community antenna TV. A wired coaxial-based network used for TV broadcasting.

CBR Committed bit rate.

CCITT Consultative Committee on International Telegraphy and Telephony. The previous name for the ITU-T before it changed in the early 1990s.

C Connector type of bayonet locking connector for coaxial cable.

Cellular Polyethylene Expanded or foam polyethylene.

Central Member The center component of the cable. It serves as a strength member or an antibuckling element to resist temperature-induced stress. The central member is comprised of steel, fiberglass, or glass-reinforced plastic.

Centralized Cabling A cable containing both fiber and copper media per article 770 of the NEC.

CEU Commercial end users. (See *SU*.)

Channel Return (Echo) The ratio of the reflected signal to the transmitted signal.

Character A symbol such as a number, letter, punctuation, or control function.

Chrominance Signal The portion of a composite video signal that contains the color information.

Circuit A communications path or network between two devices (for example, end system or switched).

Circuit Switching A network method for carrying information over a dedicated path, connecting the source and the destination through several network nodes. The connection is established over a series of physical links in which a logical channel for the connection is reserved on each of the links. Circuit switching is suitable for a continuous flow such as voice in telephony.

Circular Mil The area of a circle with one-one thousandth of an inch diameter. By knowing the circular mils of a conductor and its composition, various characteristics can be determined.

Cladding The coating on the fibers of a fiber-optic cable.

CLEC Competitive local exchange carrier (LEC). Companies in competition with the traditional exchange carriers. (See *ILEC, LEC*.)

CO Central office. Premise of a carrier service provider where customer lines (telephone lines) are multiplexed and switched to other COs.

Coating A material put on a fiber during the drawing process to protect it from handling and the environment.

Coax Cable Coaxial cable. Cable with a center copper strand, surrounded by an insulating dielectric, which in turn is covered with a grounding shield and, finally, covered by insulation.

Code The rules that specify the way data are to be presented and read.

Coil Effect The inductive effect exhibited by a spiral-wrapped shield.

Color Code A system of different colors or stripes to identify the value of electrical components or wires.

Component Video The unencoded output of a camera or videotape recorder in which the primary colors and sync pulses are transmitted down a separate cable.

Composite Video The encoded video comprised of color and sync pulses transmitted down a single cable.

Concentric Stranding A group of uninsulated wires twisted together and containing a center with subsequent layers spirally wrapped around the core.

Conduit A tube of metal or plastic through which wires or cables can be run. Used to protect the wires or cable and to make metal conduit relatively fire proof.

Connector A device designed to allow electrical signals to flow from a cable to a device or a device to a cable. A connector allows a cable to be disconnected from a device without cutting the cable.

Connector Panel A panel insert designed for use with patch panels; it contains adapters pre-installed for use when field connecting fibers or other media.

Connector Panel Module A device designed for use with patch panels for connecting connectorized fibers that can be spliced to backbone cable fibers.

Contact Durability The number of insertion and withdrawal cycles that a connector must be able to withstand and still be within the limits of specification.

Control Codes Codes that cause equipment to perform certain functions.

Control Unit The unit that directs traffic between a host computer and the I/O devices.

Convergence Common industry term for the development of procedures, protocols, and technologies to enable the merger of different applications, such as voice, video, and data, on a single transmission system or network while preserving and supporting the distinct performance requirements of each.

Copperweld® Trademark of Copperweld Steel Co. for copper-clad steel conductor.

Core The central region of an optical fiber through which light is transmitted.

Corona The ionization of gases about a conductor that results when the potential reaches a certain level.

CoS Class of service. The term that refers to the set of performance parameters that characterize the performance characteristics of a given service in relation to other services on the same network. In a DIFFServ-enabled IP network, these parameters would be included in the DS byte. (See *QoS*.)

Coupling The transfer of energy between two or more components or cables in a circuit. (See *Adapter*.)

Coverage How well a metal shield covers the underlying surface, measured in percentage.

CPC Customer premises communication.

CPE Customer premises (or provided) equipment. Data communication equipment (hardware and software) located at the customer's site.

CPU Central processing unit.

Crimp To secure the buffer tubing of a fiber cable to the tabs on a connector, or to secure the shield of a coaxial cable to a connector.

Crimping Dies Often referred to as die sets or die inserts, these are tools containing jaws that when brought together, produce the desirable deformation of the connectors that have been placed between them.

Crimping Tool A hand-held device that is used to crimp a connector.

Cross Talk The transfer of an unwanted signal from one circuit to another by magnetic or electromagnetic coupling.

CRT Cathode ray tube. Used as the display for television, oscilloscopes, and monitors.

CS Data Access™ Cross-Connect System. A device that allows DOS channels to be individually routed and reconfigured.

CSA Carrier service area.

CSU Carrier service unit. Equipment at the user end that provides interface between the user and the communications network (the T1 lines).

CTI Computer telephone integration. The term for the procedures that support the coordination of a LAN-based data system and a PBX-based voice system so that the voice and data functions occur at an ACD agent's desktop for an incoming call. (See *Tap.*)

Current A movement of electrons through a circuit. The force that carries the energy through the circuit. The unit of current is the ampere (A).

Current Loop A two-wire transmit/receive loop.

Current Rating The maximum continuous current recommended for an electrical device under given conditions, such as given temperature. The rating can also be defined in peak current for a given time.

Cut-Through Resistance A test to determine the ability of a material to withstand the application of sharp edges without being cut.

D1 A component digital video recording format that conforms to the CCIR-601 standard. Records on 19-mm magnetic tape. The term is often used incorrectly to indicate component digital video.

D2 A composite digital video recording format. Records on 19-mm magnetic tape.

D3 A composite video recording format that records on ½ in (1.27 mm) magnetic tape.

Database A large collection of information into computer files.

Data Communications The process of transferring information from one data processing device to another. The transfer is usually in digital form.

dB (Decibel) A comparison of signal levels based on the base 10 logarithm functions. The comparison can be between power levels, voltage levels, or current levels. A 50% power loss is a –3 dB loss. A measurement of 1/10 Bel. One milliwatt (electrical) = 0 dBm (optical).

DBS Direct broadcast satellite services.

DCE Data communications equipment. Equipment that provides signal conversion between the DTE and the network.

DC Resistance See *Resistance.*

DDS Digital data service. Private line digital service that provides digital communication circuits.

DEC Digital Equipment Company, one of the leading manufacturers of microcomputer-related hardware and software.

DECNET Digital Equipment Corporation proprietary architecture.

Delay Delay can refer to the actual transmission time of data, the waiting (queuing) time, time in a buffer (queuing delay, QD), time it takes for the data to travel between any networks, time that it takes data to travel end to end in network nodes (propagation delay, PD), processing time (packetization, depacketization, protocol processing, or coding), or the time for data to be switched through a switch or router (switching time).

Delay Line A transmission or equivalent device designed to delay a signal for a specific time.

Device An input/output unit such as a printer, PC, or other workstation.

DHCP Dynamic host configuration protocol. A TCP/IP protocol that provides static and dynamic address management.

Diagnostic Procedures used to test and troubleshoot a system.

Dielectric The insulation material between the center conductor of a coaxial cable and the shield. Nonmagnetic and therefore nonconductive to electrical current. Glass fibers are considered dielectric. A dielectric cable or material contains no metal.

Dielectric Breakdown   The point at which a dielectric material becomes conductive. Normally the failure of an insulator due to excessive voltage.

Dielectric Constant   The ability of an insulating material to hold an electric charge when placed between two conductors. The number is based on the dielectric of vacuum, which is 1.

Dielectric Heating   The heating of an insulating material when placed in a radio frequency field, causing internal losses.

Dielectric Loss   The power dissipated in a dielectric material when exposed to a high-frequency field.

Dielectric Strength   The voltage that an insulating material can withstand before it breaks down.

Dielectric Withstanding Voltage (DWV)   The maximum potential difference that a dielectric can withstand without failure.

Digital Signal   An electrical signal that possesses two distinct levels corresponding to 0s and 1s.

Discontinuity   Rated interconnection: a broken connection, open circuit or loss of connection characteristics, or transient phenomenon. Unacceptable variations of current or voltage or short-term interruptions.

Disk   An electronic storage device for digital information.

Dispersion   The cause of bandwidth limitations in a fiber. Dispersion causes a widening of input pulses along the length of the fiber. The three types are (1) modal dispersion, caused by differential optical path lengths in multimode fiber; (2) chromatic dispersion, caused by differential delay of various wavelengths of light in a waveguide material; and (3) waveguide dispersion, caused by light traveling in both core and cladding materials in single-mode fiber.

Distortion   Any unwanted deviation of the information signal. Distortion may be a change of amplitude, phase, frequency, or shape of a signal.

DLC   Digital loop carrier. Equipment outside the CO, usually in a neighborhood, that concentrates analog loop lines, digitizes and multiplexes calls, and multiplexes the calls to the CO.

DLCI   Data link connector identifier. The number of the virtual circuit corresponding to a particular connection between two destinations. This number is used as part of the frame relay number.

DMT   Discrete multitone. DSL technology using digital signal processors to divide the signal into 256 subchannels.

Downstream   The transmission direction from the CO to the customer's premises.

Down Time   The time when all or part of a system is out of operation and not available to the user.

DRAM   Dynamic random access memory. Memory used to store data in PCs and other devices.

Driver   A hardware device used to connect to many loads, or a memory resident program used to control a hardware device.

Drop   The connection on a circuit; a cable from a distribution panel to a faceplate.

Drop Cable   In CATV, the signal cable from the distribution box to the residence.

DS-0   Digital signal level zero. Specifications for a 64-kbps channel. Originally defined as digital transport of a single voice connection, it can be used to transport any 64-kbps digital signal. No physical specifications were ever standardized for the transport of a single DS-0 channel. (See *DS-1, T1.*)

DS-1   Digital signal level 1. Electrical interface for digital transmission at the rate of 1.544 Mbps containing 24 64-kbps DS-0 circuits. The physical interface first defined to carry a DS-1 electrical signal is known as T1 and operates over a pair of unshielded twisted-pair (UTP) wires. DS-1 has subsequently been specified to operate over a wide variety of media such as microwave and fiber optics. The terms *DS-1* and *T-1* are often (incorrectly) used interchangeably.

DS-3   Digital signal level 3. Electrical interface for digital transmission at the rate of 44.736 Mbps. It can simultaneously support 672 DS-0 circuits (28 DS-1 signals). Originally specified to operate over coaxial cable, it is now frequently implemented over microwave.

DSL   Digital subscriber line. The general name for several specifications supporting digital customer local loops.

DSLAM   DSL access multiplexer. The central office half of a DSL system. It concentrates multiple subscribers and provides access to the appropriate data services such as ISPs. (See *ATU-Central Office, ATU-Remote.*)

DSP   Digital signal processor. A microprocessor at the physical level that performs the digital modulation, demodulation, noise filtering, sampling, encoding, decoding, and/or frequency conversion of an analog signal.

DSU   Data service unit. Equipment at the user end of an access service that acts as a telephony-based interface between low-rate (56 kbps) services and higher bit rate circuits. (See *CSU*.)

DTE   Data terminal equipment. Equipment, such as a computer or terminal, that provides data in the form of digital signals for transmission.

DVDDS   Digital video and data delivery system.

DWDM   Dense wave division multiplexing. A popular name for WDM, when the number of simultaneous light wavelengths being multiplexed is very large, generally 32 or more. For example, 32 light wavelengths, each carrying 2.5 Gbps (STS-48, STM-16), multiplexed via a DWDM would result in a single optical stream operating at 2.5 Gbps for the wavelength but supporting 80 Gbps on the aggregate fiber (2.5 Gbps × 32).

E-1 A   Wideband digital interface operating at 2.048 Mbps, as defined by the ITU recommendations G.703 and G.704. Generally available outside North America.

Earth   British terminology for zero reference point (ground).

Echo   The return of a signal on a line usually due to an impedance mismatch. Also, the reflected signal in a radar system.

EIA/TIA   Electronics Industry Association/Telecommunications Industry Association. This organization provides standards for the data communications industry to ensure uniformity of the interface between DTEs and DCEs.

Elastomer   Any material that can return to its original dimensions after being distorted.

Electromagnetic   The electrical and magnetic fields caused by electron motion through conductors.

Electromagnetic Compatibility (EMC)   The ability of an electronic device to operate within its intended environment without its performance being affected by EMI and without generating EMI that might affect other equipment.

Electromagnetic Coupling   The transferring of energy by a varying magnetic field, called induction coupling.

Electromagnetic Interference   Unwanted electromagnetic interference that causes undesirable responses, degrading performance or causing malfunction of electronic equipment.

Electron Volt   The energy gained by an electron falling through an electric field produced by a 1-V differential.

Electrostatic   Electricity at rest, as an electric charge, called static electricity.

Electrostatic Coupling   The transfer of energy by way of a changing electrostatic field through capacitive coupling.

ELFEXT   Equal level far-end cross talk.

Elongation   The increase in length of a conductor or cable due to longitudinal stress.

EMF   Electromotive force (voltage).

EMI   Electromagnetic interference.

EMS   Element management system. A management system that provides functions in the element management layer.

Energy   The ability to do work.

Energy Dissipation   Loss of energy in a system due to the conversion of unwanted heat loss. Dissipation of electrical energy occurs when a current passes through a resistance.

ENG   Electronic news gathering.

Entrance facility   An entrance to a building for both public and private network service cables, including the entrance point at the building wall and continuing to the entrance space or room.

EPR   Ethylene-propylene-diene monomer rubber. A material with excellent electrical insulating properties.

Equilay   More than one layer of helically laid wire with the length of each layer the same.

Equipment Room   A centralized space for communication equipment that serves the occupants of the building. An equipment room is usually considered distinct from a communication closet because of the nature or complexity of the equipment.

Ethernet   A bus-based LAN protocol using CMA/CA, officially known as IEEE 802.3 standard. Originally designed to operate up to 100 Mbps on shared broadcast media, new approaches have emerged that enhance the operational characteristics and speed of Ethernet-based LANs. The system operates so that any station may transmit a data packet so long as the bus is idle. If two stations transmit simultaneously, they detect the foreign signal (not their signal) and "back off" for a later retry.

ETP   Abbreviation for a copper-refining process called electrolytic tough pitch. A process that produces a 99.5% pure copper, resulting in high conductivity.

ETSI   European Telecommunications Standards Institute. A European organization that produces technical standards in the area of telecommunications.

EU   European Union. Formally known as the European Commission.

EV   Electron volt.

Fanout   Multifiber cable constructed in the tight buffering design. Designed to ease connectorization and rugged applications for intra- or interbuilding requirements.

Farad   A unit of capacitance that will store 1 C of energy. One ampere per second will raise the voltage 1 V across a 1-F capacitor.

Far-End Cross Talk (FEXT)   The coupling of received signals onto another received signal of adjacent pairs.

FAS   Fire Alarm and Signal Cable, CSA (Canadian Standards Association).

Fast Ethernet   A 100-Mbps Ethernet.

Fault   An open, break, or an intentional short in a circuit.

FCC   Federal Communications Commission. A U.S. government agency that is responsible for setting standards for electronic transmission and their enforcement for all electrical communication in the United States.

FDDI   Fiber distributed data interface. A LAN token ring standard for 100-Mbps fiber-optic local area networks.

FDI   Feeder distribution interface. Points where cable bundles from the telephony switch use drop lines extended out to the customer premises.

FDM   Frequency division multiplexing. A communications transmission method that is used to transmit multiple signals over a single channel. Each signal transmitted is modulated at a different frequency and multiple signals are transmitted over the channel.

Feedback   Energy that is extracted from one point in a circuit and reflected to another. Positive feedback, called regeneration, causes an increase in gain and possible oscillations. Negative feedback, called degeneration, produces a reduction in gain and reduced signal level and signal stability.

Feeder Cable   In a CATV system, the transmission cable from the head end to the trunk amplifier. Also called trunk cable.

Ferrous   Comprised of or containing iron. A ferrous metal has magnetic characteristics; that is, it is attracted to a magnet field and can be magnetized.

Ferrule   A mechanical fixture, generally a rigid tube, used to confine and align the stripped end of a fiber.

FEXT   Far-end cross talk. A category of cross talk that occurs at the remote end of a link.

Fiber   Thin filaments of glass or plastic. An optical waveguide consisting of a core and a cladding that is capable of carrying information in the form of light.

Fiber Bend Radius   The radius that a fiber can bend without the danger of breaking or attenuation.

Fiber Distributed Data Interface (FDDI)   See *FDDI*.

Fiber Optics   Transmission media comprised of glass or plastic fibers that are used to pass light signals rather than electrical signals in communication systems. Fiber-optic cables have a much greater bandwidth than copper cables.

File   A collection of data records.

File Attributes   Access rights assigned to each file.

Flame Resistant   The ability of a material not to fuel a flame once the heat source is removed.

Flexibility   The ability of a cable to bend in a short radius.

Flex Life   The ability of a cable to bend many times without breaking.

Floating   A circuit that has no reference to ground.

Fluorocopolymer   Generic term for PVDF.

FM Modulation   Frequency modulation of a carrier. A method whereby the carrier frequency is shifted in rhythm with the information frequency.

Foam Polyethylene   See *Cellular Polyethylene*.

FOTP   Fiber-optic transmission system. A transmission system that utilizes fiber cabling.

Frame Relay   A medium-speed, connection-oriented data packet switching technology that provides reliable and efficient packet delivery over virtual circuits (VCs). It supports access speeds up to 1.544 Mbps (T1) or 2.048 Mbps (E1) in Europe. The basic transport unit, called a frame, can be up to 4096 octets and carry both routing and user information. Much of the network layer functionality is handled by the frame relay link layer, including signaling, acquiring network services, and interface management.

Frequency   The number of times that a periodic action occurs in one second, measured in Hertz.

Frequency Power   Normally 50- or 60-Hz power.

Fresnel Reflection Losses   Losses at the input and output of optical fiber due to differences in the refraction index between the glass core and the cladding glass.

FRF   Frame relay forum.

FTP   File transfer protocol. A TCP/IP standard protocol that allows a user on one host to access and transfer files to and from another host over a network.

FTTC   Fiber to the curb. A residential access technology to provide broadband service over fiber-optic cabling directly to the home. Connectivity supports data rates of up to 622 Mbps downstream (to the customer) and up to 155 Mbps upstream (customer to the network). (See *FTTH, HFC*.)

FTTH   Fiber to the home. A residential access technology to provide broadband services over fiber optics directly to the home. Connectivity supports data rates up to 622 Mbps downstream and up to 155 Mbps upstream.

Fusion   The operation of joining fibers together by fusion or heat.

Fusion Splice   A permanent connection produced by the application of sufficient localized heat to melt and fuse the ends of the optical fiber, forming a continuous single fiber.

Gage (Gauge)   The diameter of a wire. (See *American Wire Gauge, AWG*.)

Gain   The increase of current, voltage, or power. Measured in decibels.

Gateway   A device that acts as a translator between networks that use different protocols.

Geosol®   A solderable, extra tough film insulation developed by Belden for use in geophysical cables and miniature cables.

Gigahertz (GHz)   A unit of frequency equal to 1 billion cycles per second.

GND   Ground.

Graded Index   Fiber design in which the refractive index of the core is lower toward the outside of the fiber core and increases toward the center of the core. This bends the rays inward and allows them to travel faster in the lower index of the refraction region, allowing higher bandwidth capability for multimode fiber transmission.

Grommet   Plastic or rubber edging used around cable and pigtail entrance to holes such as termination centers.

Ground Loop   An undesirable circuit condition in which interference is created by ground currents when grounds are connected at more than one point.

Ground Potential   Potential of the earth. A terminal, chassis, or conductor that is used as the reference point for other potentials in a circuit.

GUI   Graphical user interface.

H.323   Specifications for the transmission of voice and video over LANs that do not provide guaranteed QoS. Defines protocols and entities such as H.323 terminals, gateways, and gatekeepers.

H.323 Terminal   A device that has implemented the H.323 protocol in support of voice or video over a LAN. A PC with H.323 software, a sound card, and a microphone; a telephone with a H.323 adapter; an IP telephone; or an IP PBX. (See *Gateway*.)

Half Duplex   In half-duplex communication, the terminal transmits and receives data in separate, consecutive transmissions.

Handshaking   Part of the communication protocol in data transfer.

HDB3   High-density bipolar three zeros substation. A line-coding technique used to accommodate the ones density requirements for E1 lines.

HDSL   High-bit-rate digital subscriber line. An alternate specification to T1 for the physical implementation of DS-1 1.544-Mbps transmission over two pairs of twisted copper wire. HDSL offers 1.544-Mbps transmission over four wires that do not require the special engineering of T1.

HDSL2   High-bit-rate digital subscriber line 2. A modified version of HDSL, under development, that can support full DS-1 1.544-Mbps transmission over a single pair of twisted copper wires instead of two twisted pairs.

Headroom   The amount by which a cable ACR exceeds 10 dB. The TIA/EIA 568B standard requires a minimum of 10 dB of ACR in Category 5 certification.

Hertz (Hz) Frequency measurement; 1 cycle per second equals 1 Hertz. Hertz is usually applied to define measurement of signals other than pulses. For example, 100 cycles per second of a sinewave equals 100 Hz, whereas 100 digital pulses per second equals 100 bits per second (bps).

Hexadecimal A digital numbering system used for coding, consisting of 16 bits. The bits are numbered 0 through 9 followed by A, B, C, D, E, and F.

HF High frequency.

HFC Hybrid fiber coax. A residential access technology to provide broadband services over the cable TV infrastructure (fiber to curb and coax in the residence). Current connectivity supplies data rates of 18 Mbps downstream (to customer) and 280 kbps upstream (customer to network). (See *FTTC, FTTH.*)

Horizontal Cabling Communication cabling that provides connectivity between the horizontal cross connect and the work area outlet. Horizontal cabling consists of transmission media, the work area outlet, and horizontal cross connect.

Horizontal Cross Connect (HC) A cross connect of horizontal cabling to other cabling (for example, to horizontal backbone or equipment).

Host Computer The controlling or central computer in a data communication network.

HTML Hyper Text Markup Language. An authoring software used on the World Wide Web. Basically, HTML is ASCII with HTML commands.

HTU-C HDSL terminal unit—central. The module at the CO or central site end of an HDSL connection. Also known as a network termination unit (NTU).

HTU-R HDSL terminal unit—remote. The module at the customer premise end of an HDSL connection. Also known as a network termination unit (NTU).

HUM A term used to describe 50- or 60-Hz noise present in some communications equipment. Hum is usually the result of electromagnetic coupling for a 50- or 60-Hz source or the result of poor filtering of the power supply rectifier.

Hybrid Cable A fiber-optic cable containing two or more different types of fiber, such as multimode and single mode.

Hypalon® A DuPont trade name for synthetic rubber used as an insulating and jacketing material for wire and cable.

IAD Integrated access device. A device designed to provide interworking functions between non-ATM equipment and an ATM network. The device may be standalone or integrated into the equipment.

ICEA Insulated Cable Engineers Association.

ICX Inner-exchange carrier. A public switching network carrier that provides access in conjunction with local exchange carriers (LECs). With the implementation of the Telecommunication Act of 1996, the distinctions between ICXs and LECs are declining.

IDC Insulation displacement connection. A wire connection device.

IDF Intermediate distribution frame. A cable distribution frame.

IDSL ISDN-like digital subscriber line. A variation of ISDN for digital data access in which the two B-channels are combined into a single 128-kbps dedicated connection or the two B-channels and the D-channel are combined into a single 144-kbps connection. In either case, no signaling is supported and the connection is dedicated. (See *DSL.*)

IEEE Institute of Electrical and Electronic Engineers.

IETF Internet Engineering Task Force. The primary working body developing TCP/IP standards for the Internet.

IF Intermediate frequency. Usually the frequency in a receiver between the RF frequency and the audio frequency.

ILEC Incumbent LEC. The local carrier exchanges that were created from the local Bell operating companies as a result of the breakup and divestiture of the Bell system in 1984 as well as other local exchange carriers that were in existence at the time. (See *LEC.*)

Impedance The total opposition that a circuit offers to current at a given frequency. Impedance is the vector addition of reactance and resistance in a circuit. The symbol Z is used to denote impedance. Impedance is given in ohms with a phase angle.

$$Z^2 = R^2 + (X_L^2 + X_C^2) \text{ or } Z\angle\theta.$$

Impedance Characteristics   In a transmission line of infinite length, the ratio of current and voltage at the point the voltage is applied, or the impedance that makes a transmission line seem infinitely long when connected across the line's output terminals.

Impedance, High   Generally, above 25 kΩ.

Impedance, Low   Generally, below 25 kΩ.

Impedance Match   A condition whereby the impedance of a transmission line is matched to a circuit, a device, or another transmission line.

Impedance Match SubA section of a transmission line, called a matching sub, cut to match the impedance of a load. May also be called a matching stub.

Impedance Matching Transformer   A transformer designed to match the impedance of one circuit or line to another circuit or line.

Index-Matching Fluid   A fluid or gel with an index close to or equal to that of glass that reduces reflections caused by refractive-index differences between glass and air.

Index of Refraction   The ratio of light velocity in a vacuum to its velocity in a given transmission medium.

Index Refraction   The relationship of light velocity in a vacuum to its velocity in a given transmission medium.

Inductance   The property of a wire that produces a counter EMF that opposes the current in the wire. A wire may be coiled into a "coil" or inductor to increase the inductance. Inductance is measured in Henries (H).

Induction   The phenomenon of a magnetic field, voltage, or electrostatic charge being produced in an object by lines of force from another object.

Induction Heating   Heating a conducting material by placing it in a rapidly changing magnetic field. The current induced in the material by the changing magnetic field produces I^2R losses that generate heat.

INI   Intelligent network interface.

Injection Laser Diode   A junction-diode laser in which the lasing (light emission) occurs at the junction of the n-type and the p-type semiconductor materials.

Input   The power or signal that is applied to the input terminal of a piece of electrical equipment.

Input Device   A device in a data processing system from which data may be entered into the system.

Insertion Loss   A measure of the attenuation of a component or cable by determining its loss before inserting the component or cable into a system.

Insulation   A material having dielectric properties, of high resistance and low leakage, used to separate conductors in electrical environments such as cables, transformers, and conductors on printed circuit boards.

Insulation Crimp   The area of the connector that is formed around the insulation of the cable.

Insulation Displacement Contact (IDC)   A termination technique whereby an insulated wire is forced into a slot on a contact that pierces the insulation and connects the contact and the conductor.

Insulation Grip   The ability of some connectors to hold fast to both a wire insulation and the conductor.

Insulation Resistance   The electrical resistance, in ohms, between two conductors separated by insulation.

Interface   The connection between equipment or the equipment that interfaces one system to another or the faces of a multicontact conductor that face each other when the conductor is assembled.

Interference   An electrical or electromagnetic disturbance that causes an undesirable response in electronic equipment.

Intermediate Cross Connect (IC)   A secondary cross connect in the backbone cabling used to mechanically terminate and administer backbone cabling between the main cross connect and the horizontal cross connect.

Internet   A worldwide connection of computers and LANs.

Internet Packet Exchange (IPX)   One of the Internet data transmission protocols used by NetWare.

Interoperability   The ability of equipment from different vendors to communicate using standardized protocol.

Intrabuilding Backbone   The portion of the backbone cabling between buildings.

Inverted Latch   A modular jack that is mounted on a printed circuit board with its contacts on the bottom and the latch on the top.

Invertible Jack   A jack that can be mounted on a printed circuit board so that the input can be from the top or the bottom.

Ionization   The formation of ions. An ion is an atom that has gained or lost an electron. An ion that has gained an electron is considered negative, and an atom that has lost an electron is considered positive.

Ionization Voltage   The potential at which a material such as air gives up an electron and becomes ionized.

IP   Internet protocol. An open network protocol used for Internet packet delivery.

IP Switching   The marketing name to an approach merging IP Layer 3 routing with the basic functions of ATM Layer 2 cell switching onto a routerlike platform. The term was coined by the Ipselon network, which is no longer in existence. It was the forerunner for the development of Cisco's Tag Switching and IBM's MPLS.

IPX   Internet Packet Exchange. A LAN communications protocol used to move data between server and workstation programs running on different network nodes.

I^2R Formula for power in watts. I is current and R is resistance.

IR Drop   The designation of a voltage drop in terms of current and voltage.

IRS   Ignition radiation suppression.

ISA   Industrial Standards Association.

ISDN   Integrated Services Digital Network. A CCITT standard that is primarily concerned with the control of voice and data.

ISO   International Standards Organization. An international standards organization that has developed an OSI communication model for data communication protocol. The OSI model is widely accepted in the international community.

Isolation   The ability of a component or circuit to reject interference, usually measured in decibels.

ISP   Internet service provider. A vendor that provides direct access to the Internet.

ITU   International Telecommunication Union. A United Nations treaty-based standards and specification body, formerly known as CCITT, whose published recommendations cover a wide spectrum of areas involving definition of terms, basic principles and characteristics, protocol design, description of models, and other specifications. Part of the International Telecommunication Union (ITU) founded in 1948 to promote telephony and telegraph issues.

Jack   A connecting device into which a plug may be inserted to form an electrical contact.

Jacket   The outer protecting covering of a wire or cable. Usually the jacket provides insulation as well as protection from the environment.

Jumper   Cable that has connections installed at both ends. (See *cable assembly*.)

kbps   kilobits per second (kbps). One kilobit is usually taken as 1024 bits per second.

KEV   1000 electron volts.

Kevlar™   See *Aramid Yarn*.

Kilo   1000 (10^3).

Kilometer   One thousand m or 3281 ft. The meter is the unit of measurement of distance in most of the world. The meter is the standard of length in fiber optics. (1 foot = 0.3048 m.)

KPSI   A unit of tensile strength expressed in thousands of pounds per square inch.

KV   Kilovolt; (10^3) volts.

KVA   Kilovolt ampere.

KW   Kilowatt.

LAN   See *Local Area Network*.

Lane   LAN emulation. Typically used in LANE over ATM.

LAP-F   Linked access protocol—frame relay. Q-922 framing.

Laser   A coherent source of light with a narrow beam and a narrow spectral bandwidth of about 2 mm. Light amplification by stimulated emission of radiation. A device that produces coherent light with a narrow range of wavelengths.

Laser Diode   An electro-optical device that produces coherent light with a narrow range of wavelengths, typically centered around 780 nm, 1320 nm, or 1550 nm. Lasers with wavelengths centered around 780nm are referred to as compact disk (CD) lasers.

Last Mile   The local loop, a distance of about 3 miles or km. This is the distance between a local telephone company office and the customer's premises.

LATA   Local access and transportation area. A geographically defined telecommunications area within which a local carrier can provide communication services.

Lay   The length measured along wires or cables required for a complete turn around the axis. In a twisted-pair cable, the lay is the distance that it takes for the two wires to make a complete turn around each other.

Lay Direction   The direction of the twist or wrap of wires or cables. The lay direction can be either right or left, clockwise or counterclockwise.

Layer   OSI reference model. Each layer performs certain tasks to move information between the sender and the receiver. Protocols within each layer define the task for that layer within networks.

Lead In   The cable that provides the signal path to or from the antenna to a receiver or transmitter.

Leakage   The undesirable passage of current through or across an insulating material.

Leased Line   A telephone line reserved for the leasing party.

LEC   Local exchange carrier. A company that provides local communication services (IntraLATA).

LED   A light emitting diode. A device that is used as a readout matrix, as an indicating light, and as light transfer to electrical signal in a fiber-optic system.

Level   A measure of the difference between a quantity or value and an established reference.

LF   Low frequency.

Light Emitting Diode (LED)   A device used in a transmitter to convert information from electric to optical energy.

Limpness   The ability of a cable to conform to a surface, as with ribbon-floor cable.

Line Drop   A voltage or power loss occurring along a transmission line due to resistance, reactance, or leakage along the line.

Line Level   Refers to the output voltage of electronic equipment, expressed in dB (fdB).

Line Voltage   The potential value of a voltage existing on a power line or supply line.

Link   A communication cable between two communications devices, excluding the equipment connectors, such as between a transmitter (source) and receiver (detector).

LLOPE   Linear low density polyethylene jacketing.

LMDS   Local multipoint distribution system. A cell-based transmission system for fixed-position wireless transmission connectivity that could replace wire-based loops. LMDS operates in the 28 to 31 GHz range.

Load   Drawing excessive power from a line. Usually results in voltage drop, power loss, and heating.

Local Area Network (LAN)   A geographically limited communications network intended for the local transportation of data, voice, and video. A network that interconnects PCs, terminals, workstations, servers, printers, and other peripherals over short distances (usually within the same floor or building). Various LAN standards are used, with Ethernet the most common. Other standards are the token bus and the token ring protocol.

Local Exchange   The central or local exchange where the subscriber lines terminate.

Local Loop   The distance between the CO and the customer's premises.

Loop   A unidirectional closed signal path connecting input/output devices.

Loose Tube Cable   Type of cable design in which the fiber cable is encased in buffer tubing for fiber protection.

Loss   Power or signal loss without any useful work.

Low Frequencies   A band of frequencies between 30 kHz and 300 kHz, as designated by the FCC.

Luminance Signal   The portion of the composite video television signal that represents the black and white signal or brightness.

M   Mutual inductance. The inductance that two devices, such as coils, couple back and forth between each other. $L_m = L_1 + L_2 \pm 2L_M$.

MA   One one-thousandth of an ampere. 10^{-3} A.

MAC   Media-specific access control. Protocol for controlling access to the data link layer 2.

Main Cross Connect (MC)   The centralized portion of a backbone cable used to mechanically terminate and administer the backbone cabling, thereby providing connectivity between entrance facilities, equipment rooms, horizontal cross connects, and intermediate cross connects.

Mainframe   Mainframe computer. The host computer.

MAP   Manufacturing automation protocol/technical office protocol.

Mate   To join to connectors.

Matte Finished PVC   A specially formatted PVC that looks and feels like rubber.

MATV   Master antenna television.

Mbps   Megabits per second. One megabit is considered to be 1,048,576 bits per second.

MDB   Management database. A database of managed objects used by SNMP to provide network management information and device control.

MDF   Main distribution frame.

MDPE   Medium-density polyethylene, a type of plastic material used for cable jacketing.

Mechanical Splicing   The joining of two fibers by mechanical means (fusion splicing or connectors) to enable a continuous signal. The CamSplice™ is an example of mechanical splice.

Mega   One millionth (10^{-6}).

Megahertz (MHz)   Unit of frequency equal to 1 million Hertz (1 million cycles per second).

Metric Prefixes   A group of symbols that are used to indicate powers of 10. For example, 1000 or 10^3 is written 1 k. See Tables G-1 and G-2.

Table G-1   Metric length table

Measurements of Length		
1,000 microns (m)	=	1 millimeter (mm)
10 millimeters	=	1 centimeter (cm)
10 decimeters	=	1 meter (m)
10 meters	=	1 dekameter (dkm)
10 dekameters	=	1 hectometer (hm)
10 hectometers	=	1 kilometer (km)

Table G-2   Metric weight table

Measurements of Weight		
10 milligrams (mg)	=	1 centigram (cg)
10 centigrams	=	1 decigram (dg)
10 decigrams	=	1 gram (g)
10 grams	=	1 dekagram (dkg)
10 dekagrams	=	1 hectogram (hg)
10 hectograms	=	1 kilogram (kg)

MFD   Microfarad (10^{-12} farads).

Mho   The unit of conductance (G) equal to the reciprocal of resistance. This term is obsolete and has been replaced by the susceptance (S).

Micrometer   One-millionth of a meter (10^{-6} meter). For example, 50-μm fiber.

Micron (μm)   One-millionth of an inch (10^{-6} inch).

MICRON Systems, Inc.   A leading supplier of data communications equipment.

Microphonics   Noise caused by mechanical excitation of components in a system, or electrically induced noise.

Microprocessor   A single chip that contains the processor of a computer.

Mil   A unit of length equal to one-thousandth of an inch.

Mili (m)   Prefix for one-thousandth (10^{-3}).

Mini-Bundle Cable   Siecor cable in which the buffer tube contains three or more fibers.

Mode   A term used to describe a light path through a fiber, as in multimode or singlemode.

Mode Field Density   The parameters for defining the diameter of the light-guiding region or core of a single-mode fiber.

Mode Field Diameter   The diameter of one mode of light propagating in a single-mode fiber. The mode field diameter replaces core diameter as the practical parameter in a single-mode fiber.

Modem   Modulator/demodulator. A physical device that performs conversion between analog (frequency waveform) signals and digital pulses (binary voltage pulses that represent bits). A modem is attached to a computer either externally or internally to allow access to the analog local subscriber loop of the public telephone network.

Modified Modular Jack (MMJ)   A six-wire modular jack with the locking tab on the right side used in DEC wiring systems.

Modulation   Coding of information onto the carrier frequency. This includes amplitude, frequency, or phase modulation techniques.

Monitor   The TV-like screen of a workstation.

Monofilament   A single-strand filament.

MPEG2   Motion Picture Experts Group. The second generation of a group that is developing ISO standards for full motion video.

MSDSL   Multirate SDSL.

MSO   Multiple-system operator. A cable company that operates more than one cable system.

MTSO   Mobile telephone switching office. A generic name for a main cellular switching center that supports multiple base stations.

Multichannel   Cable on which more than one channel of information can be transported.

Multifiber Cable   An optical cable that contains two or more fibers, each of which provides a separate information channel.

Multimode Fiber   An optical waveguide in which light travels in multiple modes.

Multiplex   To put two or more signals onto a single channel.

Multiplexing   Combining two or more signals, which can be recovered, into a single bit stream.

Multipoint Line   A communication line or circuit that interconnects several stations or terminals.

Multiuser Outlet   A communication outlet used to serve more than one work area, such as an open system application.

MUX   Multiplexer. A multiplexing system that can simultaneously transmit more than one signal over single transmission media.

mV   One-thousandth of a volt (10^{-3} V).

MVL   Multiple virtual lines. A new local access technology developed by Paradyne. Designed and optimized for multiple concurrent services for residential and small business markets. MVL transforms a single copper wire loop into multiple virtual lines supporting multiple and independent services.

mW   One-thousandth of a watt (10^{-3} W).

Mylar®   DuPont trademark for polyethylene terephtalate polyester film.

N×64   Describes a continuous bit stream to an application at the N×64 kbps rate.

N-Type Connector   A threaded connector for connecting coaxial cable.

Nano   One-billionth (10^{-9}).

Nanometer (nm)   A unit of measurement equal to one-billionth of a meter (10^{-9} meters).

NAP   Network access provider. The provider of a physical network that permits connection of service subscribers to NSPs.

National Electrical Code   A publication of the national Fire Protection Association (NFPA), which outlines codes for all electrical power and communication wire and cable installation in the United States.

NBR   Butadiane-acrylonitrile copolymer rubber. A synthetic rubber material with good chemical and oil resistance.

NDIS   Network driver interface specifications. Used for all communications with network adapters. NDIS works primarily with LAN managers and allows multiple protocol stacks to share a single NIC.

Near-End Cross Talk (NEXT)   The introduction of transmit signals on adjacent pairs.

NEBS   Network Equipment Building System. A set of requirements for the reliability and usability of equipment, established by Bellcore.

NEC   National Electrical Code. Defines the building restrictions for cable, such as flammability requirements.

NEMA   National Electronics Manufacturers Association.

Neoprene   A synthetic rubber with good resistance to oil, chemicals, and flame.

Network   The interconnection of computer systems, PCs, workstations, etc.

Network Address   A hexadecimal number used to identify a network cabling system.

NEXT   Near-end cross talk. Cross talk in which the interfering signal is traveling in the opposite direction as the desired signal.

Nibble   One-half bite (four bits).

NIC   Network interface card. The circuit board or other hardware that provides the interface between a DTE and a network.

NID   Network interface device. A device that connects a local loop to the customer's premises and includes a demarcation point.

NMS   Network management system. A Windows-based program that is responsible for managing a network. Implements functions at the network management layer using a network management protocol, such as SNMP.

Node   A connection or switching point on a network.

Noise   Unwanted electrical signals that interfere with data transmission. Noise can be internal or external to the system.

Nomex®   DuPont trademark for a temperature-resistant, flame-resistant nylon.

NSN   Network service node. The point in a network at which service is provided.

NSP   Network service provider. A vendor, such as an ISP, local telephone company, CLEC, or corporate LAN, that provides network services to subscribers.

NTSC   National Television Standards Commission. Organization that formulated the standards for U.S. television, also used in several other countries. NTSC employs 525 lines per frame, 30 frames per second, and 59.94 fields per second.

NTU   Network termination unit. Equipment at the customer's premises that terminates a network access point.

Numerical Aperture (NA)   The number that expresses the light-gathering power of a fiber or the angular acceptance for a fiber. It is approximately the sine of half-angle of the acceptance core. $NA^2 = n_1^2 + n_2^2$ where n_1 and n_2 are, respectively, the refraction index of the core and the cladding.

Nylon   An abrasion resistant thermoplastic material with good chemical resistance.

Ocn   Optical carrier level n signal. The fundamental transmission rate of SONET (Synchronous Optical NETwork). For example, OC3 represents a transmission rate of about 155 Mbps.

Octal   An 8-bit numbering system consisting of numerals 0 through 7 that can be used for a digital code.

OFHC   Oxygen-free high-conductivity copper, which must have over 99.95% copper content and conductivity of 101% compared with standard copper.

Ohm   The unit of electrical resistance (Ω), named after Ohmsted, who first proved resistance in wire.

Ohm's Law   The formula for the relationship among resistance, current, and voltage in a circuit. $E = I \times R$, where E is voltage in volts, I is current in amperes, and R is resistance in ohms.

Optical Fiber   Glass fiber strands in a fiber-optic cable along which light energy is transported.

Optical Waveguide Fiber   A transparent filament of high refractive index core and low reflective index cladding that transmits light.

OTDR   Optical time domain reflectometer. Used to measure light fiber attenuation in fiber and across a connector.

OSI   Open Systems Interconnections. The OSI Reference Model is a seven-layer network architecture model of data communication protocols developed by OSI and ITU.

Output   The useful signal power that is delivered by a circuit or device.

Output Device   A device in a data processing system from which data may be received from the system.

Ozone   Extremely active form of oxygen, occurring around electrical discharges and present in the atmosphere in small amounts. In large amounts it can break down certain types of rubber insulation.

Packet   A discrete unit of data bits transmitted over a network.

PAL (Phase Alternate Line)   The European color television system featuring 625 lines per frame, 25 frames, and 50 fields per second.

Panel Stop   The feature on a connector that prevents the connector from falling through the panel.

Parallel Circuit   A circuit in which components are connected across each other (in parallel). The voltage is equal across every component in a parallel circuit.

Parallel Digital   Digital information that is transmitted in parallel.

Patch Cord   A flexible cable terminated at both ends with a plug. Used for interconnecting circuits on a patch panel.

PBX   Private branch exchange. A user's branch exchange of telephone switching equipment dedicated to one customer. A PBX connects private telephones to each other and to the public dial network.

PCB   Printed circuit board. An insulated board (usually fiberglass), on which copper clad forms the conductive wires for an electronic circuit.

PE   Polyethylene. A type of plastic material used in the manufacture of plastic material used for cable jacketing.

Peak   The maximum instantaneous voltage of a voltage waveform.

Perfusing   Fusing with a low current to clean the fiber ends as a procedure for fusion splicing.

Periodicity   The uniformly spaced cable impedance variations that result in reflections of a signal. These impedance variations cause standing waves that affect signals transmitted along the line.

Phase   An angular relationship between waves, or the relationship between voltage and current in a circuit.

Phase Modulation   A type of signal modulation in which the phase of the carrier signal is shifted by the modulation signal.

Phase Shift   A change in the phase relationship between two or more electrical quantities.

Photodetector (Receiver)   A device that converts light into electrical energy. Silicon photo diodes and avalanche photodiodes are used in these devices.

Pick-Up   Any device that is capable of transforming a measurable quantity of sound, heat, pressure, stress, etc., into a relative electrical signal.

Pico   One-trillionth (10^{-12}).

Picofarad 10^{-12} Farad. Abbreviated pF.

Pigtail   Fiber-optic cable that has connectors installed on one end. (See *Cable Assembly.*)

Pin Diode   A photodetector used to convert optical signals to electrical signals.

Plastic   High polymeric substances, comprised of both natural and synthetic products that are capable of flowing and being molded under heat. Plastic materials can be reheated and recycled.

Plasticizer   A chemical added to plastics to make them soft and flexible.

Plenum   A compartment, chamber, or vent in which one or more air ducts are connected to form part of an air distribution system.

Plenum Cable   A cable that has been UL approved as having high heat resistance and low toxic- and smoke-producing properties that allow it to be installed in heating air ducts.

Plug   A connecting device that is inserted into a jack. It is usually connected to a cable.

Point-to-Point Connection   A communication circuit that links two points.

Polybutadiene   A type of synthetic rubber used to blend with other synthetic rubbers to improve their properties.

Polyethylene   A thermoplastic material with excellent electrical properties.

Polymer   A substance made of many molecules or chemical units. The term is often used to describe rubber or plastics.

Polyolefin   Any of the polymers and copolymers of the ethylene family.

Polyurethane   A class of polymers with good abrasion and solvent resistance.

Polyvinyl chloride   A general-purpose plastic used for wiring insulation and cable jacketing.

POP   Point of presence. An IXC's or NSP's equivalent of a CO.

Port   A computer interface point, usually a connection where information can be input or output to communicate with another device.

POTS   Plain old telephone service. Standard telephone over the PSTN, with analog bandwidth of less than 4 kHz.

POTS Splitter   A device that filters out the DSL signal and allows the POTS frequency to pass.

Potting   Sealing or filling with a substance to bar moisture.

Power   The amount of work per unit of time. In electrical circuits, expressed in watts. In other areas, expressed as horsepower.

Power Loss   The difference between the power delivered to a device or cable and the power delivered from that device.

Power Ratio   The ratio of the power into a device or cable and the power out of the device or cable. Normally expressed in decibels.

Precision Video   Video having very close electrical tolerances, requiring cables with close tolerances in impedance, velocity of propagation, signal attenuation, and structural return loss. Required in live broadcast network studios.

Prefusing   A low-current electric arc used to clean the fiber end prior to fusion splicing.

Printed Circuit Board   A metal-coated board that is etched to form the electrical conductors for a circuit.

Propagation Delay   Time for an electronic circuit or device to transmit a signal from its input to its output.

Propagation Delay Time   The time between the application of a digital pulse into a circuit or device and the corresponding change in the output waveform. The delay is usually different for a positive pulse and a negative pulse.

Protocol   A set of rules that govern (establish, maintain, and manage) end-to-end or node-to-node communication between two peer (at the same level) entities across a network.

Prototype   A module suitable for evaluation of form, design, and performance.

PSC   Public Service Commission. State-level regulators of public telephone companies in the United States.

PSELNEXT   Power sum equal level near-end cross talk.

Pseudo Random NRZ   Non-return to zero signals that do not return to zero volts. NRZ signals are used in computer applications.

PSNEXT   Power sum near-end cross talk.

PSTN   Public switched telephone network. A generic name for the worldwide telephone network. (See *POTS, Public Networks, VPN.*)

PTO   Public telephone operator.

PTSS   Passive transmission subsystem.

PTT   Postal telephone & telegraph. Provider of access services. A government agency in many countries.

Public Network   A telephone company such as AT&T owned by stockholders.

Pulse   A voltage that changes abruptly from one value to another and back in a finite length of time.

PUR   Polyurethane material used in the manufacture of a type of jacketing material.

PVC   Permanent virtual circuit. A virtual connection established administratively, including in network supporting ATM, frame relay, and X.25.

PVC   Polyvinyl chloride. Material used in the manufacture of a type of jacketing material.

PVDF   Polyvinyldifluoride, a plastic material used for jacketing of cabling. Often used in plenum cabling.

QAM   Quadrature amplitude modulation. A modulation technique that uses variations of signal amplitude.

QoS   Quality of service. The set of performance standards that characterize the performance characteristics of a given service in numeric, absolute terms.

Quad   A four-conductor cable, sometimes called a "star" quad.

RADSL   Rate-active digital subscriber line. A variation of ADSL that adapts data rates to the condition and capabilities of the local loop circuits. (See *DSL.*)

RBOC   Regional Bell operating company. Local regional telephone companies that resulted from the breakup of AT&T.

Receiver   An electronic package that converts optical signals to electrical signals. Can also be a device to receive electromagnetic radio waves and amplify them for a useful purpose.

Reflectance   The ratio of the power reflected to the incident power at the connector junction or other component. Reflectance is measured in negative decibel values. The larger the decibel value of reflectance, the better the connector. For example, a -40-dB connector would have one-half the reflection of a -37-dB connector.

Refractive Index   See *Index of Refraction.*

Registered Jack (RJ)   Represents the code given to jacks utilized in interconnections across a system. These codes are part of the USOC. Examples are RJ1,1 which is a six-position modular jack or plug, and RJ45, which is known as the eight-position modular jack or plug.

Repeater   A device that consists of a transmitter and a receiver or transceiver, used to regenerate a signal to increase the system transfer range.

Resistance   The property of an electrical cicuit that limits the current.

Return Loss   See *Reflectance.*

RF   Radio frequency.

RFI   Radio frequency interference. Levels of computer-generated electrical noise regulated by the FCC.

RH   Relative humidity.

Right-Angle Jack   A jack that has entry to the side when mounted to a printed circuit board.

Ring   A type of connection for a computer system. A network in which a unidirectional signal path is a closed loop.

Ring Topology   A network configuration that connects all nodes in a ringlike configuration.

Rise Time   The measurement of the time it takes a pulse to rise from 10% of maximum value to 90% of maximum value.

Riser   Pathways between floors for indoor cables, normally a vertical shaft or space that must meet the NEC fire code.

Riser Application   Used for indoor cables that pass between floors. Normally a vertical shaft or space.

ROM   Read-only memory.

Rope Strand   A conductor comprised of groups of twisted strands.

Router   A device that connects LANs by dynamically routing data according to network layer 2.

RS-232   An EIA-recommended standard for a computer cable terminal connection that specifies the voltages and electrical character of the pin signals. The 25-pin connection is the most used cable connection.

RSVP   Resource reSerVation setup Protocol. Provides priority data transmission based on reservation protocol.

RT   Remote terminal. RTs are intermediate points close to the customer's premises to improve service reliability. The local loop can be terminated at an RT.

RTU   Remote terminal unit. A device installed at the customer premises that connects to the local loop. Also referred to as the ATU-R.

Rubber   A general term used to describe wiring insulations, such as natural or synthetic rubber, Hypalon®, neoprene, and butyl rubber.

SBR   The most commonly used type of synthetic rubber. A copolymer of styrene and butadiene.

Scattering   A property of glass that causes light to deflect from the fiber and contribute to signal loss.

Screened Twisted Pair (SCTP)   A cable type of 120 Ω impedance, used in Category 5 applications.

SDH   Synchronous Digital Hierarchy. Based in part on SONET, SDH is an ITU standard for the interworking of ANSI and ITU transmission techniques.

SDSL   Symmetric or single-pair digital subscriber line. An emerging specification for supporting full DS-1 1.544-Mbps and ITU transmission techniques.

Semiconductor   A material processing method used to develop wires with electrical conductivity somewhere between that of conductors and insulators. In the wiring industry a semiconductor is made by adding carbon particles to an insulator. This term is different and is not to be confused with that used for silicon and germanium for making diodes and transistors.

Separator   A layer of insulating material added between a conductor and insulator of a wire or cable, or between the various components of a cable to improve flexibility and aid in stripping.

Serial Digital Information (SDI)   Digital information transmitted in a serial form.

Serial Transmission   A transmission in which the bits are transmitted one at a time and in sequence.

Series Circuit   An electrical circuit in which the components are end to end. The current of a series circuit is the same in each component, and the voltage is divided across the components with respect to their resistance values. The total resistance of a series circuit is $R_T = R_1 + R_2 + R_3$ ....

Serve Shield   A metallic shield consisting of several strands of wire.

Sheath   A metal or insulating cover of the wires in a cable to protect them from wear, moisture, and stress.

Shield Coverage   The optical percentage of a cable actually covered by a shield.

Shield Effectiveness   The ability of a shield to block interference of unwanted signals, not to be confused with shield coverage.

Shield/Shielding Cable   A conducting envelope, composed of metal strands of foil, that encloses a wire or groups of wires so constructed that substantially every point of the surface of the underlying insulation is at ground potential or at some predetermined potential with respect to ground.

Shield/Shielding Circuit   The metal shielding surrounding a connector to prevent interference, interaction, or current leakage. Such shielding protects a circuit against cross talk.

SI Units   Table G-3 lists the most common SI (standards international) units.

Table G-3   Basic Units of SI System

Quantity	Unit	SI Symbol
Length	Meter	M
Mass	Kilogram	kg
Time	Second	s
Current	Ampere	A
Light Intensity	Candela	cd
Molecular Substance	Mole	mol
Thermodynamic Temperature	Kelvin	K
Angle	Radians	rad
Capacitance	Farad	F
Conductance	Siemen	S
Electric Charge	Coulomb	C
Flux Density	Tesla	T
Force	Newton	N
Power	Watts	W
Inductance	Henry	H
Magnetic Flux	Weber	Wb
Magnetic Force	Ampere-turns	a°t
Resistance (Electric)	Ohm	Ω
Potential	Volts	V
Velocity	Meters per second	m/s
Temperature	Degrees Celsius	°C
Work (Energy)	Joule	J

Signal   Information conveyed through a communication system.

Signal-to-Noise Ratio   A ratio comparison of the desired signal to unwanted signals on the line or in the system.

Silicone   A General Electric trademark for a material made of silicon and oxygen. Silicon can be liquid or thermosetting and has a high heat tolerance.

Single Ended   An unbalanced line or circuit grounded at one end.

Single-Mode Fiber   An optical waveguide (or fiber) in which the signal travels in one mode. The source is usually a light emitting diode (LED), typically 8.3 to 9.5 µm.

Sinusoidal   A wave or other action varying in time as the sine function. Alternating current is sinusoidal.

SIP   SMDS Interface Protocol. A three-layer protocol implemented in SMDS networks.

Skew Ray   A ray that does not intersect the fiber axis. A light ray that enters the fiber at a very high angle.

SKI   Symmetric key infrastructure. Generates a one-time password as a challenge response through an algorithm that uses changing factors like time and card usage counts to generate a random password. Solves the password theft problem since passwords are no longer valid over multiple uses.

Skin Effect   The tendency of an alternating current to travel on the surface of a conductor at high frequencies.

SMDS   Switched multimegabit digital service. A packet switching service serving LANs.

SNA   System network architecture. A description of the logical structure and protocols that transmit information and control the operation of an IBM network.

Snake Cable   The name given to individually shielded or individually shielded and jacketed multiple audio cables.

SNMP   Simple Network Management Protocol. Protocol for open network management.

SNR   Signal-to-noise ratio, used interchangeably with ACR. (See *ACR*.)

Software   A computer program(s) that controls the operation of logic circuits to perform certain functions.

SOHO   Small office/home office.

SONET   Synchronous Optical NETwork. An ANSI standard for the transmission of digital data over optical networks.

Source   The means used to convert an electrical information-carrying signal to a corresponding optical signal for transmission by fiber. The source is usually an LED or laser.

Spectral Bandwidth   The difference between wavelengths at which the radial intensity is one-half that of the peak intensity.

Spectrum   Frequencies that exist in a continuous range and have common characteristics such as the radio spectrum, the light spectrum, and the electromagnetic spectrum.

Splice Closure   A container used to house, organize, and protect cabling within cabling trays. Typically used in outside plant environments.

Splice Tray   A container used to organize and protect spliced fibers.

Splicing   The permanent joining of fiber ends to identical or similar fibers without the use of a connector. (See *Fusion Splice, Mechanical Splicing*.)

Standing Wave   A stationary pattern of waves produced by two waves of the same frequency traveling in the opposite direction on the same transmission line. Reflected energy from an impedance mismatch in line termination causes these high and low voltage patterns.

Standing Wave Ratio (SWR)   The ratio of the maximum to the minimum value of current or voltage along a transmission line.

Star   Coupler optical component that allows emulation of a bus topology in fiber-optic systems.

Star Quad   A four-conductor spiraled microphone cable used to increase noise rejection.

Static Charge   An electrical charge on a body caused by the production of ions on the body, usually produced by friction.

Station   Any setup to perform a computer operation (for example, a PC station).

Stay Cord   A high-tensile-strength component of a cable used to anchor the cable end at its termination.

Step-Index Fiber   Optical fiber that has an abrupt (step) change of its refractive index due to a core and cladding that have different indices of refraction. Typically used for single-mode operation.

Step Insulation   The process of applying two layers of insulation on a cable. Used in shielded network cables so that the outer insulation can be removed and terminated in an RJ-type connector.

STM   Synchronous time modulation.

STP   Shielded twisted pair.

Strain Gauge   A device used to measure the change in dimension of a material when stress is applied.

Strand   A single uninsulated wire.

Stranded Conductor   A conductor comprised of several strands of wire.

Structural Return Loss (SRL)   The measurement in decibels of signal reflection in a cable.

STS-1   Synchronous Transport Signal-1. The fundamental SONET standard for transmission over fiber at 51.84 Mbps.

SU   Server unit. The end user at the customer premises.

Surge   A temporary and relatively large transient voltage occurring on a line.

SWC   Service wire center.

Sweep Test   Testing the frequency response or attenuation over a range of frequencies. This is accomplished by generating a constant voltage sweep over a range of frequencies and observing a graph of the output of the cable or device being tested.

Switch   A device that is responsible for switching traffic based on real or virtual circuits. It is also common for a switch to perform traffic management and control functions such as admission control, congestion control, and signaling.

There are various switch architectures, which can be classified according to different characteristics (buffering, switch matrix, interconnection design, multiplexing, and switch control).

2BIQ Two binary, one quaternary. A line-coding technique that compresses two binary bits into the time frame of one as a four-level code.

10BASE-T The technical name for Ethernet twisted-pair wiring.

T1 A TDM digital channel carrier that operates at the rate of 1.544 Mbps over two twisted-pair wires. Often confused with DS-1, the specification for framing and transmitting a 1.544-Mbps bit stream consisting of twenty-four 64-kbps DS-0 channels. DS-1 was originally specified to be transmitted over T1 wiring but has subsequently been specified on other types of media as well. As a result, the terms *T1* and *DS-1* are often used interchangeably.

T3 The informal name for any TDM digital channel that operates at 44.73 Mbps. T3 can multiplex 28 T1 signals and is synchronous with DS-3.

Tap A connection to the main transmission line in a LAN. For example TAP1 or TSA1.

TCP/TP Transmission Control Protocol/Internet Protocol. The dominant protocol suite in the World Wide Web.

TC-PAM Trellis-coded post amplitude modulated.

TDM Time-domain modulation.

TDR Time-domain reflectometer.

Teflon® DuPont trademark for fluorocarbon resins.

Teleco Telephone company.

Telecommunication (Communication) Closet An enclosed space for housing communication equipment, cabling termination, and cross connects between the backbone and horizontal cabling.

Teleprocessing The handling of data processing information utilizing communication equipment.

TELEX A network of printers connected over an international public network.

Telnet A virtual terminal protocol in the Internet suite of protocols. Allows the user of one host computer to log into a remote host computer.

Tensile The amount of axial load (longitudinal stress) required to break or pull a wire from a connector.

Tensile Strength The greatest longitudinal stress that a substance or union can bear without tearing or pulling apart. In a crimped termination it is the greatest longitudinal stress that a connector can bear without the wire separating from the crimped connector.

Terminal Block A terminal strip with connections through and from which cables can connect.

Termination The connecting of a connector at the end of a cable or the termination of the cable in its characteristic impedance.

Thermal Rating The temperature range in which a material can perform its function without undue degradation.

Thermoplastic A material that will flow or distort under heat or pressure.

Thermosetting A material that will not flow or distort under heat or pressure.

Throughput A performance metric that gives the aggregate maximum bandwidth capacity (in bps) of a system (such as a switch, link, network) or the actual bandwidth a user is using for transmission, measured over an observed interval of time. Useful throughput refers to the net throughput achieved by application, excluding any overhead due to control information included in the packet's retransmission due to lost packets or other losses.

TIA Telecommunication Industry Association.

TIA/EIA 568-A A commercial building wiring standard for voice and data communications. The specification was jointly developed by the Telecommunications Industry Association and the Electronics Industries Association.

Tight-Buffered Cable Type of cable construction whereby each fiber is tightly buffered by a protective thermoplastic coating to a diameter of 900 μm, resulting in high tensile strength, durability, ease of handling, and ease of connection.

TIMS Two binary, one quaternary. A line-coding technique that compresses two binary bits of data into one time state as a four-level code.

Tinsel A type of conductor comprised of tiny threads, used for small-sized cables that must be highly flexible.

TNC A threaded type of connector used on coaxial cable.

Token Ring A type of connection for a LAN computer system in which a code (token) is passed around the ring to a certain station. The station with the token can receive the information from the ring.

TP-PD Twisted pair—physical dependent.

Training Time   The time necessary to train an operator or user to operate a piece of equipment.

Transducer   A device for converting one type of energy into another type. For example, a photo coupler converts light energy into electrical energy.

Transfer Impedance   Transfer impedance relates the current on one surface of a shield to the voltage induced on the opposite surface of the shield. Another surface transfer impedance rating is used to rate the effectiveness of interfering signals both internal and external. The higher the transfer impedance rating of a shield, the more effective the shield.

Transmitter   An electronic package that converts electrical signals to optical signals, or electrical signal to electromagnetic waves.

Triboelectric Noise   Noise induced in a cable due to changes in capacitance between the conductor and the shield as the cable is flexed.

Trunk Cable   See *Feeder Cable*.

Turn Key   A contractual agreement in which one party designs and installs a system and turns it over, fully operational, to another party who will operate the system.

Twinaxial Cable   A shielded cable consisting of one or more twisted-pair cables.

Twin Lead   A transmission line with two conductors separated by insulation. The line impedance, usually 300 Ω, is determined by the size of the conductors and their spacing.

Twisted Pair   Two copper wires twisted to cancel out electromagnetic induction. Many twisted pairs may be combined in a single bundle to form a cable that is covered by a tough outer jacket. Each pair is specifically color coded.

UAWG   Universal ADSL, Working Group.

UBS   Universal serial bus.

UHF   Ultra-high frequency, ranging from 300 to 3000 MHz.

UL   Underwriters Laboratories, Inc. A U.S.-based company that tests the standards of products for safety and reliability. Manufacturers pay to have their equipment tested to acquire a UL stamp of approval.

Unbalanced Line   A transmission line, such as a coaxial cable, on which the voltages on the two conductors are unequal with respect to ground.

UNI   User-to-network interface.

Unilay   A conductor with more than one layer of helically laid wires with the length and direction of the lay the same for all layers.

Universal Service Order Code (USOC)   A set of designations for carrier to customer interfaces. (See *Registered Jack*.)

Upstream   Typically refers to the transmission speed from the customer premises toward the telephone network.

URL   Uniform resource locator. An Internet standard addressing protocol for location and access resources.

UTP   Unshielded Twisted Pair. Two copper wires twisted around each other to reduce induced noise. Two or more pair of these wires are usually enclosed in an outer jacket to form a cable. Twisted pair are usually formed with 24- or 26-gauge wires.

VA   Volt-Ampere. A designation of power.

VC   Virtual circuit. A logical connection or packet-switching mechanism established between two devices at the beginning of a transmission.

VCSEL   Vertical cavity surface emitting laser. A type of laser used for transmitting light into a fiber.

VDSL   Very high-bit-rate digital subscriber line. An emerging specification established between two devices at the beginning of a transmission.

Velocity of Propagation (VP)   The transmission velocity of energy along a transmission line with respect to transmission velocity in space. Usually expressed in a percentage.

Vertical Jack   A jack that utilizes only top entry.

VLAN   Virtual LAN. Workstations on different LANs can be connected using VLAN tagging.

VLF   Very low frequency, 10 Hz to 30 kHz.

VLSI   Very large scale integration.

VOD   Video on demand. A service allowing many users to request the same video simultaneously.

VOIP   Voice over IP. A general term for several techniques for the transmission of voice over an IP-based router network.

Voltage   The measure of electromotive force that forces current through a circuit.

Voltage Drop   The voltage developed across a component by a current through resistance, reactance, or impedance.

Voltage Rating   The highest voltage that a material or component can withstand without breaking down.

VPN   Virtual private network. In a VPN, resources such as bandwidth and buffer space are provided, on demand, to users in such a way that users view a certain partition of the network as a private network. The advantage of VPNs over private networks is that VPNs allow a dynamic use of public network resources while retaining much of the controls, operational characteristics, and security of a private network at a much lower personal equipment service cost. VPN services are available from PSTN carriers for voice service and IPS carriers for Internet support.

VSWR   Voltage standing wave ratio. The ratio of the transferred voltage signal to the reflected voltage signal on a transmission line.

VW-1   A flammability rating established by Underwriters Laboratories for wires and cables that pass a specially designed vertical flame test, formally designated FR-1.

Wall Thickness   The thickness of insulation or jacket of a wire or cable.

WAN   Wide area network. A network that spans a large geographic area, such as a city.

Wavelength   The distance between two crests of an electromagnetic or light signal measured in meters, micrometers, or nanometers. The symbol for wavelength is γ and is formulated as follows:

$$\gamma = (300 \times 10^{-6} / \text{frequency in Hz}) \text{ meters.}$$

WIPO   World Intellectual Property Organization.

Wire   A bare or insulated conductor.

Work Area   An area in which personnel operate data processing equipment.

Work Area Telecommunications Outlet   A connecting device located in a work area in which the horizontal cabling terminates and provides connectivity for work area patch cords.

Work Stations   Areas in which there is data processing equipment such as terminals and printers.

WWW   World Wide Web. An interactive, multimedia, Internet-based application that has become extremely popular and useful due to its highly integrated presentation and distribution of information and ease and speed of accessing services and information around the world. HTTP is the protocol that forms the basis of the WWW. Information in audio, text, and video is prepared on the Web by a language called HTML, which creates the actual hyperlinks.

X The symbol for reactance in ohms.

X_C Capacitive reactance. The reactance or opposition that a capacitor offers to an AC current flow.

xDSL   The general name for a variety of digital subscriber line (DSL) technologies. (See *ADSL, HDSL, IDSL, VDSL*.)

X_L Inductive reactance. The reactance or opposition that an inductor offers to an AC current.

Z   The symbol for impedance in ohms.

Zero-Dispersion Wavelength   Wavelength at which the chromic dispersion of an optical fiber is zero. This occurs when the waveguide dispersion cancels out material dispersion.

VENDOR INFORMATION

The following list of vendors is a small sampling of the thousands of manufacturers and distributors that handle communication products and materials. The authors suggest, for further reference, one of the search engines on the Internet, or the *Telephone Industry Directory* by Phillips' Publishing, Inc., 7811 Montrose Road, Potomac, Maryland, 20854, (800)722-9120. This directory contains thousands of vendor addresses and phone numbers.

Another source is the Internet under Telecommunications or Communications.

Allied Electronics-Sub Digitech, Inc.
(lineman's tools, test equipment, and wire and cable)
401 East 8th Street
Fort Worth, TX 76102
(800)433-5700
www.allied.avnet.com

Amdahl Communications, Inc.
(data network controllers, data sets, data test equipment, data transmission equipment, multiplexers, and repeaters)
2330 Millrace Court
Mississauga, Ontario L5N IW2
Canada

American Cable Corp
(coaxial, plastic insulated jacked, plenum, plug/connector-ended, and switchboard cable; key telephone connectors; and modular retractable cords)
711 Cooper Street
Beverly, NJ 08010
(416)821-9900

AMP, Inc.
(premise cabling solutions)
USA products center
(800)522-6752

AMP, Inc.
(cable assemblies and connectors, fiber-optic components)
www.amp.com

AMP Products Corp
(interconnection products, including closures, fiber optics mechanical strippers, cross-connect systems, and under carpet wiring and accessories)
450 West Swedesford Road
Berwyn, PA 19312
(215)647-1000
(800)522-6752

Anixter Brothers, Inc.
(adapters; wiring systems; products for voice, video, data; and power application. Provides manufacturing and refurbishment of telecommunication products; outside plant materials for telephone construction; customer premises, transmission, and central office equipment; earth station satellites; and local area network equipment. Distributes transmission and plant equipment products for AT&T, Belden, BICC, Brand-Rex, Cablec, IBM, Keptel, Porta, Raychem, Rome. Siecor, Telco Systems, Thermo Electric, 3M, and TRW.)
Skokie, IL
(708)677-9480

AT&T Network Systems
(network switching, transmission, and outside plant products)
475 South Street
Morristown, NJ 07960
www.at&t.com

AT&T Technology Systems
(fiber-optic cable, connectorized cable, circuit boards, connectors, converters, equalizers, multiplexers, fiber-optic transmission systems, power supplies, and wire)
One Oak Way
Berkeley Heights, NJ 07922

Automatic Tool and Connector Co.
(fiber optic cable tools and accessories)
(800)524-2857

Belden Electronic Wire and Cable
(cables and assemblies, connectors, fiber-optic components, and data links)
2200 South U.S. 27
Richmond, IN 47375
(317)983-5200
www.belden.com

CIS/Suttle Apparatus
(communication cabling accessories)
www.suttleonline.com

Data Products International
(IBM-compatible telephone company management software that runs on IBM and handles accounting, capital credits, commercial billing, general plant, inventory, and toll rating applications)
P.O. Box 1176
Sugar Land, TX 77487
(713)491-7200

Hewlett-Packard
(electronic and communication test instruments)
www.hp.com

International Business Machines, Inc. (IBM)
(mainframes, PCs, premise wiring, software, local area network installation, system servicing, wiring, and cabling hardware System Product Department)
One Culver Road
Dayton, NJ 08810
(800)IBM-2468

ITT Pomona
(test and measurement equipment accessories)
(800)241-2060/Int.(909)469-2928

Kline Tools
(electrical and electronic tools)
www.klinetools.com

Nevada Western
(wire management systems and local area networking products, local area networking design and installation)
615 North Tassman Avenue
Sunnyvale, CA 94089
(408)734-2700

Pomona
(test and measurement accessories)
www.pomonaelectronics.com

ProGain, Inc.
(twisted pair wiring enhancement units)

Simpson
(electrical and electronic instruments)
853 Dundee Avenue
Elgin, IL 60120
(847)697-2260

Riser Bond Instruments
(time domain reflectometers, VOPs, and other measuring equipment)
5101 North 57th St.
Lincoln, NE 68507
www.riserbond.com

Siecor
(iber optic cabling and components, fiber optic study guide)
www.siecor.com

Tektronix
(electronics and communication measuring instruments)
www.tektronix.com

Wandel & Goltermann, Inc.
1116 Radisson St.
San Jose, CA 95130-2063
www.wg.com

INDEX

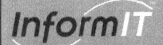

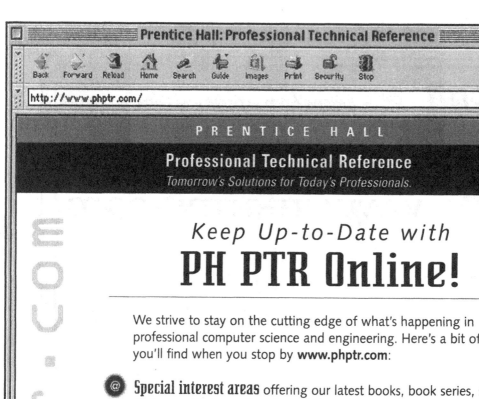